CAMBRIDGE LIBRARY COLLECTION

Books of enduring scholarly value

Life Sciences

Until the nineteenth century, the various subjects now known as the life sciences were regarded either as arcane studies which had little impact on ordinary daily life, or as a genteel hobby for the leisured classes. The increasing academic rigour and systematisation brought to the study of botany, zoology and other disciplines, and their adoption in university curricula, are reflected in the books reissued in this series.

Studies in Bird Migration

Having trained as a civil engineer and surveyor, the ornithologist William Eagle Clarke (1853–1938) established himself in his field by preparing reports on bird migration for the British Association. Focusing on the species passing through the British Isles, Clarke spent many months in various lighthouses and on remote islands. He brought all his research together in this two-volume work, first published in 1912 and illustrated with maps, weather charts and photographs of key research locations. In Volume 1, Clarke notes which species arrive in the British Isles during each season. A map shows the routes they take. He also explains how weather conditions affect avian journeys, using charts to indicate temperature changes across Europe and wind conditions over Britain. The annual movements of swallows, skylarks, rooks and other species are then discussed individually. The volume closes with Clarke's account of the month he spent at the Eddystone Lighthouse.

Cambridge University Press has long been a pioneer in the reissuing of out-of-print titles from its own backlist, producing digital reprints of books that are still sought after by scholars and students but could not be reprinted economically using traditional technology. The Cambridge Library Collection extends this activity to a wider range of books which are still of importance to researchers and professionals, either for the source material they contain, or as landmarks in the history of their academic discipline.

Drawing from the world-renowned collections in the Cambridge University Library and other partner libraries, and guided by the advice of experts in each subject area, Cambridge University Press is using state-of-the-art scanning machines in its own Printing House to capture the content of each book selected for inclusion. The files are processed to give a consistently clear, crisp image, and the books finished to the high quality standard for which the Press is recognised around the world. The latest print-on-demand technology ensures that the books will remain available indefinitely, and that orders for single or multiple copies can quickly be supplied.

The Cambridge Library Collection brings back to life books of enduring scholarly value (including out-of-copyright works originally issued by other publishers) across a wide range of disciplines in the humanities and social sciences and in science and technology.

Studies in Bird Migration

VOLUME 1

WILLIAM EAGLE CLARKE

CAMBRIDGE
UNIVERSITY PRESS

CAMBRIDGE
UNIVERSITY PRESS

University Printing House, Cambridge, CB2 8BS, United Kingdom

Published in the United States of America by Cambridge University Press, New York

Cambridge University Press is part of the University of Cambridge.
It furthers the University's mission by disseminating knowledge in the pursuit of
education, learning and research at the highest international levels of excellence.

www.cambridge.org
Information on this title: www.cambridge.org/9781108066976

© in this compilation Cambridge University Press 2014

This edition first published 1912
This digitally printed version 2014

ISBN 978-1-108-06697-6 Paperback

This book reproduces the text of the original edition. The content and language reflect
the beliefs, practices and terminology of their time, and have not been updated.

Cambridge University Press wishes to make clear that the book, unless originally published
by Cambridge, is not being republished by, in association or collaboration with, or
with the endorsement or approval of, the original publisher or its successors in title.

The original edition of this book contains a number of colour plates,
which have been reproduced in black and white. Colour versions of these
images can be found online at www.cambridge.org/9781108066976

STUDIES IN BIRD MIGRATION

I.

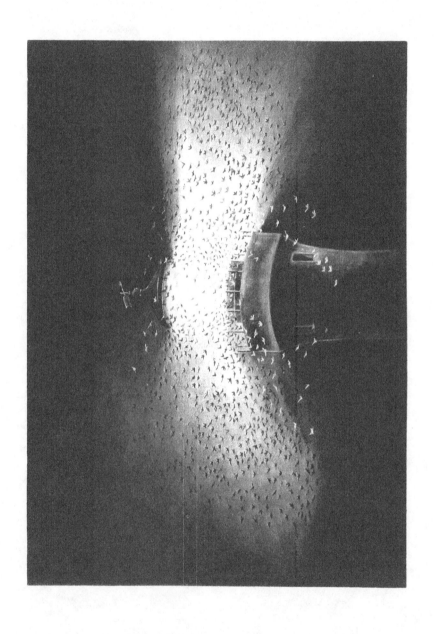

THE EDDYSTONE LANTERN

12TH OCTOBER 1901

From a painting by Marian Eagle Clarke

. . . The beacon's blaze allures
The bird of passage, till he madly strikes
Against it, and beats out his weary life.—*Tennyson.*

[*To face Frontispiece.*

STUDIES

IN

BIRD MIGRATION

BY

WILLIAM EAGLE CLARKE

Keeper of the Natural History Department, the Royal Scottish Museum

WITH MAPS, WEATHER CHARTS, AND OTHER
ILLUSTRATIONS

VOLUME I.

LONDON
GURNEY AND JACKSON
EDINBURGH: OLIVER AND BOYD
1912

STUDIES

IN

BIRD MIGRATION

WILLIAM EAGLE CLARKE

WITH MAPS, WEATHER CHARTS, AND OTHER
ILLUSTRATIONS

VOL. I.

LONDON
GURNEY AND JACKSON
EDINBURGH: OLIVER AND BOYD
1912

TO HER GRACE

THE DUCHESS OF BEDFORD

Honorary Member of the British Ornithologists' Union

IN RECOGNITION OF VALUED ASSISTANCE WHICH HAS CONTRIBUTED
MATERIALLY TO THE KNOWLEDGE OF A SUBJECT OF MUTUAL
INTEREST, THESE STUDIES ARE GRATEFULLY INSCRIBED
BY THE AUTHOR

PREFACE

THESE Studies are offered as contributions to what has ever been an attractive branch of the science of Ornithology. They are the result of many years' personal observations and researches, during which exceptional opportunities have been afforded me for acquiring special knowledge of the subject.

In the year 1880 it was my good fortune to become the intimate friend of the late John Cordeaux, who first inspired me with an interest in bird-migration that has never waned. In 1883 I was elected a member of the British Association Committee on the Migration of Birds as observed on the British and Irish Coasts. On the completion of the investigations it was my privilege to be entrusted with the preparation of the Reports, five in number, embodying the results of that great enquiry—a task which was finally accomplished in 1903.

The preparation of these Digests revealed the fact that, vast though the data were, much desirable information was still lacking. This led me to undertake a series of special investigations with a view of contributing to our knowledge on these moot points, and in the hope of adding to the lore of the subject of Bird-Migration generally. A residence of forty - seven weeks in lighthouses and in a lightship, and fourteen weeks spent

on the remote islands of St Kilda and Ushant, enabled
me to accomplish these objects.

In addition to other aspects of the subject, attention
was devoted to the relations between migrational and
meteorological phenomena.   In carrying out these particular investigations I was singularly fortunate in enlist-
ing the aid of Dr W. N. Shaw, the Director of the
Meteorological Office, who has kindly revised the study
devoted to Migration-Weather.   The series of Charts
which so usefully illustrate that study have also been
specially prepared by Dr Shaw.

Several of the studies have appeared in the pages of
The Ibis and *Reports of the British Association;* but in
each case they have either been largely supplemented by
subsequent researches, or carefully revised in the light
of further knowledge.

A number of the stations visited had either never
before been explored ornithologically, or only partially
so, and as their bird-life is of considerable interest,
a complete concise account of their feathered natives is
afforded.

During the many years devoted to this work I was
the recipient of many acts of kindness, and have been
laid under numerous obligations.   For these, gratitude
compels me to express my deep sense of the value of the
services rendered and my high appreciation of them.

To Her Grace the Duchess of Bedford, the Elder
Brethren of the Trinity House, the Commissioners for
Northern Lighthouses, to the late Mr John Bruce
of Sumburgh, and Mrs Bruce, Mr Robert Bruce, and
the MacLeod of MacLeod, my indebtedness is due
for special facilities afforded for carrying out the
investigations.

For valued assistance of a varied nature my obligations are also due to the Hon. Gladys Graham Murray, the Misses Baxter and Rintoul, Mr Henry Johnstone, Mr Coventry Dick Peddie, Mr William Evans, Dr Harvie-Brown, and Mr John Mackenzie. Nor should the various lighthouse keepers among whom my lot has been so frequently cast be forgotten, especially Mr and Mrs Wallace of Fair Isle.

In conclusion, I must express my thanks to the Publishers for the great care they have bestowed upon the production of the work.

Wm. Eagle Clarke.

Edinburgh, *May* 1912.

CONTENTS

VOLUME I

xi

CONTENTS

VOLUME II

LIST OF ILLUSTRATIONS

VOLUME I

VOLUME II

STUDIES IN BIRD-MIGRATION

CHAPTER I

SOME ANCIENT AND ANTIQUATED VIEWS

> They try their fluttering wings, and thrust themselves in air,
> But whether upward to the moon they go,
> Or dream the winter out in caves below,
> Or hawk for flies elsewhere, concerns not us to know.
>
> —DRYDEN (1631-1700).

THE seasonal movements of birds have attracted the attention and excited the curiosity of mankind for countless ages, and the allusions to them in ancient literature are of a singularly interesting nature. Not only are these references amongst the very earliest contributions to the study of natural history, but some of them must rank as the most remarkable to be found in the literature devoted to that or any other subject. It is to be hoped that some day a historian will arise who will gather together the threads and yarns spun by the old writers, and weave them into the engaging narrative which remains to be unfolded to us. Here it is only possible briefly to allude to the vast antiquity of the records, and to the extraordinary views held by some of those who have ventured to write on the subject.

Long before the Christian era, we find in the Book of Job (xxxix. 26) the words: "Doth the hawk fly by

thy wisdom, and stretch her wings toward the south?"
This is an undoubted reference to the well-known
autumnal emigration of certain birds of prey, so much
in evidence in the East, and is probably the earliest
allusion to the subject of bird-migration.

In the opening of the Third Book of the *Iliad*—
attributed to Homer, who probably lived in the twelfth
century B.C.—we find the Trojan hosts described as
coming on with noise and shouting, "like the Cranes
which flee from the coming winter and sudden rain, and
fly with clamour towards the streams of the ocean," or,
as Pope has it :—

> So when the inclement winters vex the plain
> With piercing frosts, or thick-descending rain,
> To warmer seas the Cranes embodied fly,
> With noise and order through the midway sky.

Homer also alludes to the gathering, in autumn, of
vast numbers of water birds by the Asiatic rivers.

In the fifth century B.C., the Greek poet Anacreon
welcomed, in inimitable verse, the return of the Swallow
in spring; and it is noteworthy that he assigned Egypt
as among this bird's winter retreats.   His lines (Carmen
33) are thus beautifully rendered by Moore :—

> Once in each revolving year,
> Gentle bird! we find thee here ;
> When nature wears her summer vest,
> Thou com'st to weave thy simple nest ;
> But when the chilling winter lowers,
> Again thou seek'st the genial bowers
> Of Memphis, or the shores of Nile,
> Where sunny hours for ever smile.

A century after Anacreon, the prophet Jeremiah
(viii. 7) thus refers to the advent in spring of several
species of migratory birds: "Yea, the Stork in the
heavens knoweth her appointed time; and the Turtle

[dove], and the Crane, and the Swallow observe the time of their coming."

The above are, however, mere allusions to the subject, and it was not until a century or two later, when the works of the illustrious Greek philosopher and naturalist, Aristotle (B.C. 384-322), were given to the world, that we find the migrations of birds and other animals discussed for the first time.

These remarkable contributions are to be found in the eighth book of his *Historia Animalium*, of which, fortunately, we have recently been furnished with a masterly edition in English by Prof. D'Arcy Thompson. This notable book forms one of the volumes of *The Works of Aristotle*, and was published at Oxford in 1910. From this most valuable natural history, the following interesting information has been culled : but it is not all sound, for some highly ridiculous theories are also advanced, which are, however, pardonable, as they are at the outset of the study of any great subject.

Aristotle tells us that "Some creatures can make provision against change [of season] without stirring from their ordinary haunts ; others migrate, quitting Pontus and the cold countries after the autumnal equinox to avoid approaching winter, and after the spring equinox migrating from warm lands to cool lands to avoid the coming heat. In some cases they migrate from places near at hand ; in others they may be said to come from the ends of the world, as in the case of the Crane, for these birds migrate from the steppes of Scythia to the marshlands south of Egypt, where the Nile has its source. . . . Pelicans also migrate, and fly from the Strymon to the Ister, and breed on the banks of this river. They depart in flocks, and the birds in front wait for those in the rear, owing to the fact that when the flock is passing over the intervening mountain range, the birds in the rear lose sight of their companions in the van. . . . All creatures are fatter in migrating from cold to heat than in migrating from heat to cold; thus the Quail is fatter when he emigrates in autumn than when he arrives in spring. The migration from cold countries is contemporaneous with the close of the hot season.

" The Cushat and the Rock-Dove migrate, and never winter in our country, as is the case also with the Turtle-Dove. . . . The Quail also migrates—only, by the way, a few Quails and Turtle-Doves may stay behind here and there in sunny districts. Cushats and Turtle-Doves flock together, both when they arrive and when the season for migration comes round again. . . . When the Quails come from abroad they have no leaders, but when they migrate hence, the Glottis flits along with them, as does also the Landrail and the Eared Owl. The Swan and the Lesser Goose are also migratory.

"A great number of birds also go into hiding; they do not all migrate, as is generally supposed, to warmer countries. Thus, certain birds (as the Kite and the Swallow), when they are not far off from places of this kind, in which they have their permanent abode, betake themselves thither; others that are at a distance from such places, decline the trouble of migration, and simply hide themselves where they are. Swallows, for instance, have been often found in holes, quite denuded of their feathers, and the Kite on its first emergence from torpidity has been seen to fly from out some such hiding-place. And with regard to this phenomenon of periodic torpor there is no distinction observed, whether the talons of a bird be crooked or straight; for instance, the Stork, the Ouzel, the Turtle-Dove and the Lark, all go into hiding. The case of the Turtle-Dove is the most notorious of all, for we would defy any one to assert that he has anywhere seen a Turtle-Dove in winter time; at the beginning of hiding time it is exceedingly plump, and during this period moults, but retains its plumpness. Some Cushats hide; others, instead of hiding, migrate at the same time as the Swallow. The Thrush and the Starling hide, and the birds with crooked talons, the Kite and Owl, hide for a few days."

Thus Aristotle was the propounder of the theory known as "hibernation"—a theory which survived and held sway for two thousand years, and possibly still has a place in the belief of certain credulous persons.

Aristotle was also responsible for the theory of "transmutation"; being led to this belief by the seasonal emigration of particular species and the simultaneous appearance of others. He tells us that the Erithacus (or Redbreast) and the Redstart change into one another; the former is a winter bird, the latter a summer one, and the difference between them is practically limited to the coloration of their plumage. In the same way with the Beccafico [? Garden Warbler] and the Blackcap; these change the one into the other. The Beccafico appears about autumn, and the Blackcap as soon as autumn has ended. These birds, also, differ from one another only in colour and note; that these birds, two in name, are one in reality is proved by the fact that at the period when the change is in progress

I.   A 2

each one has been seen with the change as yet incomplete.

Passing now to the earliest years of Christian era, we find that the Roman naturalist Pliny (A.D. 20-79), in his *Historia Naturalis* (lib. x. cap. 24), has something to say on the subject of bird-migration. His views, however, are mainly those of Aristotle. He informs us that the Œnanthe (Wheatear) has its stated days for retirement; at the rising of Sirius it conceals itself, and at the setting of that star comes forth from its retreat: this it does, most singular to relate, exactly on both these days. As to the Chlorion (Golden Oriole), it is not seen in winter, but comes forth about the summer solstice. Pliny states that Blackbirds, Thrushes, and Starlings take their departure to neighbouring countries, but do not lose their feathers, nor yet conceal themselves, as they are often seen in winter. The Ring Dove also takes its departure, but whither it goes is a matter of doubt. Writing on the movements of the Stork and Crane, he remarks that up to his time it has not been ascertained from what place the Storks come, or whither they go when they leave. There can be no doubt, he tells us, that like the Cranes they come from a great distance. Pliny repeats Aristotle's statements regarding the transmutation of the birds already alluded to.

There can be little doubt that the hibernation theory owes its origin, to some extent, to the fact of Swallows, Martins, and other migratory birds having been found dead in holes in banks, etc., into which they had crept for shelter from an outburst of inclement weather during their sojourn in their native lands, which is not an uncommon event. These were supposed to be the remains of individuals which had died from natural causes while concealed in their winter retreats, and some of them may have been found in a more or less featherless condition. With the writings of Pliny the ancient views on migration come to an end. The middle ages in this, as in many other cases, appear to have been a period of intellectual stagnation, and it was not until the latter half of the sixteenth century that interest in the subject seems to have revived.

In 1555, Olaus Magnus, Archbishop of Upsala, in Sweden, published a work entitled *Historia de Gentibus Septentrionalibus et Natura*, wherein he alludes (lib. xix. cap. 29, p. 673) to the seasonal disappearance of Swallows as follows :—

"Although memorable writers on many subjects of natural history have related that Swallows change their abode, that is, seek warmer countries on being greatly pressed by winter, yet in Northern waters, by the chance of a fisherman, Swallows are often drawn out in a kind of rolled-up lump, which, when about to descend into the reeds after the beginning of autumn, have bound themselves together—mouth to mouth, wing to wing, foot to foot. Moreover, it has been remarked that they at that season, their very sweet song being finished, descend, and peacefully, after the beginning of spring, fly out thence and reseek their old nests or make new ones with their natural diligence. But if that lump be drawn out by ignorant young men (for old and expert fishermen put it back) and carried to a warm place, the Swallows, loosened by the access of heat, begin to fly about, but live only a short time, giving proof that premature birth is to be guarded against. It happens also in spring-time that the birds returning to freedom, and occupying their old nests or building new ones, being overtaken by rough weather and a heavy fall of snow, all alike die, so that none are seen the whole summer about the houses or river banks, except a very few which have risen from the deeper waters, or journeying from elsewhere, are seen, when winter is wholly dispelled in May, to arrive, about to reproduce offspring for the good of nature."

The picture here given is from a photograph taken

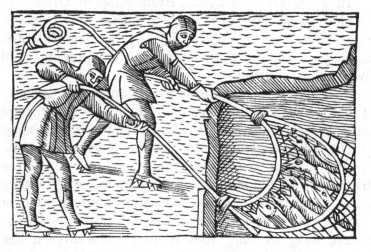

direct from Olaus Magnus's work, for the loan of which I was indebted to the late Professor Newton. It represents two fishermen standing on the edge of the ice and drawing towards them a net containing a mixed "catch" of Swallows and fishes.

This ridiculous story was accepted by a number of distinguished naturalists, including Linnæus: and John Reinhold Forster, in his edition of Kalm's *Travels into North America* (p. 140), informs us that he can reckon himself among the eye-witnesses of this "paradoxon " of natural history; he proceeds to relate how, in 1735, he saw Swallows taken from the Vistula in winter. Cuvier (1769-1832) was also a believer in the theory of submergence, for he says, " the Martin becomes torpid during winter; and that it passes the cold season under water at the bottom of marshes appears to be certain." Learned bodies such as the Royal Society of London and the French Academy have discussed this theory and published the results in their Transactions. John Hunter (1728-1793), the celebrated anatomist, made several experiments to test this theory, all of which, needless to remark, confirmed him in the opinion of the impossibility of Swallows submerging themselves or sustaining life under water. He even caught several Swallows in October and confined them in a green-house, with large tubs of water containing reeds; but found, of course, that none of the birds attempted to enter them. The Hon. Daines Barrington quotes[1] an instance of three or four Swallows (or Martins) having been found caked together in the mud at the bottom of a pond in Berkshire in February. They were carried into the kitchen, and afterwards flew about. The Rev. Dr Pye and others testify to these facts!

Swallows just ere they emigrate in the autumn have a predilection for roosting in reed beds standing in water, and this, coupled with the fact that the birds were not seen afterwards, may have led those who witnessed the act to suppose that they crept down the reeds and submerged themselves. It does not, however, justify the mythical accounts invented to bolster up the theory.

Perhaps the most extraordinary theory ever propounded regarding the migration of birds is contained

[1] *Phil. Trans.*, lxii., p. 289.

in a pamphlet published in the year 1703 by " A Person
of Learning and Piety." The title of this rare and
curious little tract is " An Essay Towards the Probable
Solution of this Question—Whence Come the Crane
and the Swallow, when they Know and Observe the
Appointed Time of their Coming?" and the "probable
solution" is that migratory birds retreat to the moon
to spend the cold season!

The author allows sixty days for the journey, and
remarks that as the moon is not a stationary body in
the heavens, "it cannot be supposed" that the birds on
embarking upon their journey "direct their course to
the moon, but rather offended by the steams of earth
do tend directly from it, and that straight line 'tis
probable they pursue, till they come so near the moon,
that she is their fairest object to draw their inclination;
for if the moon hath a motion in a month about the
earth, then at the two months' end they [the birds] find
it in the same line of direction where it was when they
began their journey; . . . therefore if they proceed in the
same straight line, they will be sure to meet the moon on
the way." The question of food and sleep on the journey
is dealt with in the same nonsensical fashion. He
says :—" As to eating, it may possibly be [*i.e.*, exist]
without in that temper of the Æther where it passeth,
which may not be apt to prey upon its spirits as our
lower nitreous air; and yet even here Bears are said to
live upon their summer fat all the winter long in
Greenland, without any new supply of food. Now we
noted before that some of those birds (and perhaps it
may be true of the rest) are very succulent and sanguine,
and so may have their provisions laid up in their very
bodies for the voyage.

" As to sleep, 'tis very probable that they are in a sleep or sweeven if not all the way between the attraction of the earth and that of the moon, to which sleep the swift acquired motion may very much contribute. . . . Now it is likely these birds being there, where they have no objects to divert them, may shut their eyes, and so swing on fast asleep, till they come where some change of air (as a middle region about the moon or earth) may by its cold awaken them. Add to this, that this sleep spares their provisions, for if (as some would have it) Cuckoos and Swallows can lie asleep half the year without eating, why cannot these in as deep a sleep as well for two months forbear it." A *résumé* of this paper, giving fuller details, appeared in the *Zoologist* for 1909, pp. 71-73.

Among the older writers on birds, Francis Willughby [1] is to be commended for his caution. Speaking of the Swallow, he remarks (p. 212): " What becomes of the Swallows in winter time, whether they fly into other countries, or lie torpid in hollow trees and the like places, neither are natural historians agreed, nor indeed can we certainly determine. To us it seems more probable that they fly away into hot countries, viz., Egypt, Ethiopia, etc., than that they lurk in hollow trees, or holes in rocks and ancient buildings, or lie in water under ice in northern countries." Willughby expresses the same views when treating of the Cuckoo (p. 98).

Later, the hibernation theory of Aristotle was resuscitated and advocated with renewed energy. Its great exponent was the Hon. Daines Barrington, whose views were expounded *in extenso* in a com-

[1] *The Ornithology of Francis Willughby*, published in the year 1678.

munication made to the Royal Society in 1772,
and entitled, "An Essay on the Periodical Appearing and Disappearing of certain birds at Different
Times of the Year."[1] Barrington's views, and his
persistent advocacy of them, do not call for special
notice, since they are well known indirectly through
the celebrated correspondence which passed between
him and Gilbert White. All ornithologists know
and regret the baneful influence they exercised over
that most excellent naturalist, whose classical work,
White's Selborne, has done more than any other
book in this country to foster the study of bird-
migration. That Gilbert White at first held orthodox
views on the subject of the migration of Swallows, etc.,
is made manifest by his letter of 4th August, 1767,
addressed to Thomas Pennant; but later he seems
to have fallen almost entirely under the influence of
Barrington, and to have become a believer in the latter's
absurd views regarding the winter torpidity of the
Swallow-kind, as did most other naturalists of that
period. Indeed, the only writer of any distinction in
the eighteenth century who adhered to sound common-
sense views on the subject was George Edwards,[2]
whose remarks are, to this day, well worthy of perusal.
Speaking of the immersion theory, "it is enough,"
writes this author, "to raise one's indignation, to see
so many vouchers from so many assertors of this
foolish and erroneous conjecture, which is not only
repugnant to reason, but to all the known laws of
nature."

[1] *Phil. Trans.*, lxii., pp. 265-326.
[2] *Natural History of Birds and Gleanings of Natural History*, published
between 1743 and 1760.

Probably nothing more remarkable can be found in the romance of natural history than some of these explanations advanced to account for the disappearance of birds in the autumn. No doubt the fact, then unknown, that most birds embark upon their emigrations at nightfall, and that hence their departure is unwitnessed, may be in some measure responsible for the astounding theories propounded to account for what was, not unnaturally, a complete mystery to men who had neither ascertained this habit by observation, nor even surmised it as a possibility; though it may well be thought strange that, among their many bold guesses, the true solution never emerged as a possible conjecture.

In looking back over the history of early opinion in relation to the appearance and disappearance of certain birds at certain seasons, it is both interesting and instructive to note that in very early times men's knowledge of the matter seems to have been scanty, but sound as far as it went, and that in later times careless observation, fancifully interpreted, brought forth "a lively principle of error," which led mankind astray for many generations, and induced even Linnæus and White—two naturalists of the first rank, and living within the period in which the scientific spirit had become powerfully operative—to lend their authority to theories now quite untenable.

CHAPTER II

SOME MODERN VIEWS [1]

WE have seen that one hundred years ago our naturalist great-grandfathers were gravely discussing whether Swallows and other migratory birds hibernated, or whether they passed the winter submerged at the bottom of lakes and rivers. To-day it is possible to give reasonable explanations for much that was formerly regarded as mysterious; but there remain problems associated with these remarkable pilgrimages which perplex the modern man of science as they did the prophet and philosopher of old.

WHAT BIRDS ARE MIGRATORY, AND WHY.—Although the vast majority of birds possess unsurpassed means of locomotion through their remarkable powers of flight, yet by no means all, nor even the majority of the known species (some 13,000 in number), are migratory. Hence the question arises: What Birds are Migrants, and Why? Those who have studied the ways of bird-life the world over, are aware that the feathered denizens of the tropics are a stay-at-home throng; while the birds

---

[1] This chapter is not offered as a contribution to the student's knowledge of the subject. It is written in the interests of readers who may not be familiar with the various phases of the phenomenon, as an introduction to the special studies which follow.

14

which are natives of the northern and temperate regions
are mostly migratory ; they are

> Intelligent of seasons, and set forth
> Their aery caravan ; high over seas
> Flying, and over lands.   —MILTON.

Indeed, with comparatively few exceptions, all those
inhabiting the northern and far southern areas are
migrants. Almost every boy and girl in our own
country knows that certain birds—the Swallow, Cuckoo,
and Fieldfare, for instance—come to us and leave us at
particular times of the year ; while those who have paid
closer attention to the bird-life of our islands are aware
that the great majority of their feathered population is
more or less migratory, coming and going with the
change of the seasons.

Why do these birds migrate? Why do they leave
their native lands and set out on long, arduous, and
dangerous pilgrimages to other climes? Is the habit of
benefit to those that practise it? These questions may
be best answered by asking yet another—namely, " What
would become of those myriads of birds which in the
summer delight in and breed amidst the solitudes of the
arctic countries, when those vast wastes which form
their feeding-grounds lie under a pall of snow or are
transformed into solid ice? What, too, would become
of certain birds which similarly make our islands their
summer home if they attempted to remain the winter
with us ? How could the Swallow and the Cuckoo, and
hosts of other birds, whose food consists of insects, and
other lowly creatures, support life during the drear
months of winter, when such food is scarce or not to be
found? The answer is an obvious one ; they would
perish from want of food, and not, as is popularly

supposed, directly from cold. They must migrate or
starve, and ages of experience have taught them that
they must quit their summer homes after their broods
are reared, and wing their way to more genial winter
quarters, mainly in the south, where food abounds.

> . . . birds that migrate from a freezing shore,
> In search of milder climes come skimming o'er :
> Instinctive tribes ! their failing food they dread,
> And beg, with timely change, their future bread.—CRABBE.

It must not be supposed, however, that the migrants
of to-day wait until their food supplies fail ; such is not
the case, for the migratory habit has become part and
parcel of their lives, through countless ages of practice,
and they usually depart from the northern lands as soon
as their offspring are able to take care of themselves.
The Swifts leave our islands, and also southern Europe,
some weeks before any failure of their food takes place—
indeed at a period when it is specially abundant.
There are imperative reasons, then, why these birds
should migrate to winter quarters more or less distant
from their summer homes. But what induces them to
quit these genial retreats and return northwards in
spring ? There are several excellent reasons. Foremost among these is the well-known passionate attach-
ment shown by birds for their native land—their true
home—in which the most fascinating period of their
lives is spent. This in itself affords the stimulus to
seek, at the appointed time, the hallowed scenes where
the all-engrossing domestic duties of the year await
them, and have for ages been performed.

> Why homeward turned thy joyful wing ?
> In a far off land I heard the voice of Spring ;
> I found myself that moment on the way ;
> My wings, my wings, they had not power to stay.—MONTGOMERY.

In this respect their patriotism, if we may so term it, is pronouncedly parochial, for the returning migrants seek the very neighbourhood (even the very field, wood, etc.) in which the old home is situated, and return to it regularly, year after year. There are many instances on record of this remarkable constancy on the part of migratory birds.

The periodic physiological changes in progress at the approach of spring in the shape of the development of the reproductive organs, with their corollary, the reviving instinct of procreation, must prove an overpowering incentive to seek the accustomed breeding haunts. These physiological developments, too, no doubt regulate the date at which the departure to the breeding grounds should take place for particular races, and will account for the late emigration from the winter home of those species which breed in the far north, where the land does not assume its summer aspect until comparatively late in the season.[1] Another important factor is that the southern and tropical regions (the main winter resorts) are not suited for a nursery for the young of the hardy northern species, and if they attempted to nest there the result would be disastrous—their race would become extinct. On the approach of the southern winter—our spring—the question of food might become an important incentive to emigration, though, personally, I do not attach much importance to it.

Thus the necessity for different summer and winter quarters, through breeding and feeding requirements, has resulted in the migratory habit becoming an integral part of the lives of a vast number of birds inhabiting the arctic and north temperate regions : while, on the other

[1] See Plate III.

hand, the teeming natives of the tropics are almost
entirely sedentary, for the simple reason that in their
fatherland—thanks to climatic conditions—food is more
or less abundant all the year round.   The migrants are
Tennyson's

> . . . happy birds that change their sky
> To build and brood, that live their lives
> From land to land.

MIGRATION IN THE SOUTHERN REGIONS.—The birds
of the south temperate and antarctic regions are also
mainly migratory, and move northwards towards and
beyond the equator on the approach of the southern
winter.   Some of them, such as Wilson's Petrel and the
Great and Sooty Shearwaters—penetrate far into the
realms of the northern hemisphere, and the two former
sometimes reach the fringe of the arctic circle.

THE ORIGIN OF THE MIGRATORY HABIT.—Dr A. R.
Wallace has offered (*Nature*, vol. x., p. 459) the
following as a likely explanation of the manner in
which the migratory habit had its origin.   He says:
"It appears to me probable that here, as in many
other cases, 'survival of the fittest' will be found to
have had a powerful influence.   Let us suppose that in
any species of migratory bird, breeding can, as a rule,
be only safely accomplished in a given area; and
further, that during a great part of the rest of the year
sufficient food cannot be obtained in that area.   It will
follow that those birds which do not leave the breeding
area at the proper season will suffer, and ultimately
become extinct; which will also be the fate of those
which do not leave the feeding area at the proper time.
Now, if we suppose that the two areas were (for some
remote ancestor of the existing species) coincident, but

by geological and climatic changes gradually diverged
from each other, we can easily understand how the habit
of incipient and partial migration at the proper seasons
would at last become hereditary, and so fixed as to be
what we term an instinct. It will probably be found
that every gradation still exists in various parts of the
world, from a complete coincidence to a complete separa-
tion of the breeding and subsistence areas; and when
the natural history of a sufficient number of species in all
parts of the world is thoroughly worked out, we may find
every link between species which never leave a restricted
area in which they breed and live the whole year round,
to those other cases in which the two areas are absolutely
separated. The actual causes that determine the exact
time, year by year, at which certain species migrate will
of course be difficult to determine."

The advances made in our knowledge on this subject
since Dr Wallace published this illuminating explanation
in 1874, all go to strengthen his views, so much so that
a good deal which was then regarded as hypothetical is
now generally accepted, while the rest may be considered
as distinctly probable.

Dr Weismann[1] has also expressed some interesting
and helpful views as to how the present distribution
of migrants in Europe came about. He is of opinion
that many of the birds which now inhabit the central
and northern regions of Europe were during the glacial
period wanting, because the climate was too severe.
They must, therefore, he considers, have come sub-
sequently from the south, and with the gradual rising
of the temperature there must have been a corresponding
steady but very gradual influx of birds into the north.

[1] *Contemporary Review*, Feb. 1879.

Just in proportion as the ice retreated, would the birds push forward the bounds of their habitat. He is careful to tell us, however, that the first migrations of birds date from an earlier period than the glacial epoch, for many species were already birds of passage before its advent.

MIGRATION ROUTES.—It is quite reasonable for the uninitiated to suppose that when journeying between these more or less widely separated seasonal homes, the migrants move indiscriminately southwards in the autumn, and, in like manner, northwards in the spring. Such is not the case. Each individual migrant of mature age has its accustomed summer and winter quarters, and follows particular and more or less devious routes to reach them. As the ways of many are identical, great and much-used fly-lines are followed. As long ago as 1846, the celebrated German ornithologist, J. F. Naumann, expressed himself thus (*Rhea*, 1846, p. 18) on this important subject : "There must," he says, "even be highway roads through the air, which are annually taken by migrants, certain spots on the earth beneath these tracts serving the travellers as stations for rest and recovery. In other tracts again migrants are either strikingly scanty or altogether absent." These "migration routes," as they are termed, have been carefully studied for Europe and northern Asia by Professor Palmen in particular, and are shown on the accompanying map (Plate II.). It will be observed that coast lines and the courses of great rivers are followed, lofty mountain ranges and wide belts of desert are traversed, and lesser or vaster expanses of sea are crossed. Hosts of migrants, even delicate Warblers and other birds of feeble wing, thread the snowy passes of the Himalayas when seeking and retreating from the

Siberian forests or the desolate tundras that border the shores of the Arctic Ocean. The waters of the North Sea are crossed, among others, by many fairy-like forms, including such frail voyagers as the diminutive Goldcrest. These are trunk routes, and the lines by which they are indicated on the maps may be regarded as their central portions; there are, however, innumerable minor ways associated with them, which it is impossible to indicate. There are also others, of a purely local nature, which are only known to observers who have been long resident in their vicinity.

In traversing these routes, Weismann[1] is of opinion that the migrants are simply following the old ways by which their ancestors originally travelled to spread themselves out towards the north. He points out, among other instances, that their present fly-lines over the Mediterranean (see map of routes) are the sites of the old land bridges (which formerly divided that great sea into a series of lesser inland seas) over which the birds used to pass on their journeys to and from the south in the earlier days of their migrations.

The journeys of many of the migrants, when travelling to their seasonal haunts, are marvellous performances, since they extend from the arctic regions to New Zealand, Tasmania, the Cape of Good Hope and Cape Horn. Some of them entail flights across wide stretches of open ocean, as in the case of the beautiful little Cuckoo (*Chalcococcyx lucidus*), which summers in New Zealand and Norfolk Island and winters in eastern Australia. The distance between New Zealand and Australia direct is about 1200 miles, and there are no resting places *en route*. The journey may be accom-

[1] *Loc. cit.*

plished *via* Norfolk and Lord Howe's Islands by three flights, two of 550 miles and one of 600 miles. Another remarkable feat is that performed by Warblers, Pipits, Shrikes, and Sandpipers, across the Himalayas, when travelling to and from Siberian and Central Asiatic summer quarters to Indian winter retreats. These birds traverse a belt of absolute desert of more than one hundred miles in width, having an elevation of over 15,000 feet, and intersected by numerous snow-capped ridges, the lowest passes of which are 18,000 feet above sea-level. These, Hume tells us,[1] present no invincible obstacles to even the tiniest and most feeble-winged migrants, such as the Warblers of the genus *Phylloscopus.*

MIGRATION BY NIGHT.—These journeys are wonderful in many ways, but how very much more wonderful do they become, when it is remembered that they are chiefly performed during the hours of darkness—indeed, almost invariably so, when a considerable expanse of sea has to be traversed! Why night travelling should be resorted to is one of the most puzzling of the many problems that are associated with the phenomena of bird-migration ; yet I think it admits of reasonable explanation. Here again, food seems to me probably to play an important part. It is well known that most of the daytime has to be devoted by birds to the search for food. Let us suppose that the traveller sets out to cross the North Sea in the daytime. What would this imply ? It would mean : (1) that the voyager must undergo a fast of many hours during its passage across the sea ; (2) it would reach our shores at night or during the early hours of the morning, and many hours must again elapse ere food

[1] Henderson and Hume, *Lahore to Yarkand*, pp. 160-1.

could be obtained. Few birds, I think, could accomplish the journey under such conditions. By crossing at night, all these difficulties are obviated. That the birds have evidently fasted for some time ere they reach our shores is manifest from an examination of their stomachs. I have examined the stomachs of many birds captured at the lantern of the Fair Isle lighthouse, and in not one of them have I found even a trace of food.

It has been suggested that birds migrate by night to escape the attacks of gulls, which often pounce down upon and destroy migrants, especially solitary ones, when they come across them at sea during the daytime. This theory is scarcely worthy of serious consideration, because such powerful birds as Cranes and Storks, among others, migrate during the hours of darkness, though surely not to escape the attention of Gulls. Then great overland passages also take place at night, in regions where there are certainly no Gulls awaiting the voyagers.

We see comparatively little of night - migration, because during clear weather it passes entirely unnoticed, even at the most favourably situated stations. Now and then the weather comes to our aid during a great migration, and we get a glimpse of how vast these movements are, and we learn something about the weather conditions which make their observation possible, some of which are noticed in the chapter on the author's experiences at the Eddystone.

DANGERS ENCOUNTERED EN ROUTE.—It must not be supposed that these journeys are free from dangers. Far from it ; the perils are many and varied. The following are a few of them. To begin with, many migrants perish at sea. Favourable weather may prevail at the point of embarkation, say in southern Norway, and induce the birds to set out to cross the sea for Britain ; for birds, like ourselves, prefer fine weather for their voyages. These favourable conditions may not extend entirely across the North Sea, and should the migrants pass into a gale before reaching our shores, which they sometimes do, many perish, and their remains are washed up in great numbers on our beaches. One of the greatest of all dangers, perhaps, is the numerous lighthouses and light-vessels on and off our own and other coasts. These, under certain conditions of weather, which will afterwards be fully described, are veritable shambles. Those who have not witnessed a " bird-night " at a light-station cannot form any conception of the appalling loss of life that takes place. I have stood on the gallery of the Eddystone lighthouse and watched the emigrants striking the great lantern and falling into the surf beneath for ten and a half hours without a break. I have stood on the deck of the Kentish Knock lightship from dusk to dawn, and seen birds falling thickly the whole time, to perish miserably, many of them being merely stunned, in the calmest of seas. On each of these occasions thousands of the migrants perished. What must the combined slaughter have been at these and neighbouring stations, or, perhaps, at many British stations, for the conditions and the movements are often wide spread? What for a whole season? Hundreds of thousands : I am convinced of this. Another danger may be alluded to, namely, the havoc wrought among the travellers by Birds of Prey. These destroyers take a heavy toll, for many follow their migrating prey and play havoc in their ranks, both during the daytime and in the dark. Owls frequently (and sometimes even the " noble " Peregrine Falcon and the " brave " little

Merlin) have been detected snatching the weary voyagers as they fly distracted round the lanterns at night. SPEED OF MIGRANTS.—The speed at which birds travel when on their migratory journeys has been much discussed. I shall give my own experiences and estimates when relating what I witnessed when in the Kentish Knock lightship, where I had many opportunities of observing birds crossing the North Sea. I do not believe that any species exceeds 100 miles per hour, and I doubt if such a speed is maintained by birds on their migrations. Gätke,[1] however, would endow them with amazing powers in this respect (as much as 207 miles—" 180 geographical miles "—per hour). The data, however, upon which his estimates are based are of the very flimsiest nature, and will not stand the tests very properly imposed by modern scientific investigators.

How ARE THE MIGRANTS GUIDED?—There still remains for consideration another question, perhaps the most interesting, certainly the most puzzling, of all. We have seen that when on their journeys, the migrants not only travel vast distances overland, but also cross pathless seas and oceans. The question is—How do they find their way? How are they guided? Here we are face to face with what is rightly regarded as one of the greatest mysteries to be found in the animal kingdom.

Many explanations have been offered. Among others it has been suggested that birds are endowed with some organ which makes them sensitive to the magnetism of the earth, and thus enables them to steer their course as a sailor steers his ship. Others look to sight and memory, both of which birds possess to a remarkable degree. It is possible that these faculties would suffice

[1] Die Vogelwarte Helgoland, 1901, p. 68.

as a guide to those that have already made the passage,
when coast lines and rivers, mountains and plains lay
before them. But of what avail would they be when a
vast expanse of sea has to be traversed? Here we
are told that during their voyages the migrants fly at
a great height,[1] sometimes at 20,000 feet, and from
such elevations the sea and land would be spread below
them like a map. To this explanation is added the
experience gained through parental guidance.

Such hypotheses, however, are quite insufficient to
explain the mystery. In the first place, we know that
most birds travel by night, and thus one fails to realise
how sight would enable them to steer a correct course
over hundreds of miles of ocean. Secondly, many
young birds migrate apart from their parents, and hence
parental guidance is not an essential factor. In support
of this latter contention, we have the remarkable evidence
afforded by the migrations performed by the young of
the Cuckoo. It is well known to naturalists that the
young Cuckoo does not leave our islands and its summer-
haunts elsewhere in the autumn, until some weeks after
the adult birds have left for their winter home ; hence
parental guidance in their case is not conceivable.
Almost everyone is aware that the parent Cuckoo
deposits its eggs in the nests of other birds, and that the
young are reared by foster parents at the expense of
their own offspring, which the infant Cuckoo soon ejects
from the nest. Now, these foster parents mainly belong
to species which do not quit our islands, or, if so, do

[1] It is possible, however, that a limit is imposed by temperature to the
height at which birds can travel during their migrations. We must
remember, too, that they wing their way through the air at a high rate of
speed, whereby the effect of cold is considerably intensified. (See *Nature*,
vol. xix., p. 481.)

not journey far ; others certainly do perform considerable migrations. It matters not, however, whether these foster parents are eminently migratory, for they do not accompany the emigrant Cuckoos on the journey to their winter retreats, which lie in the equatorial regions, or to the south of them. It is highly probable that the young of many birds go off on their own account as soon as they are able to shift for themselves, especially those which compose the first families of species which are double-brooded. Not only do these in many instances migrate apart from their parents, but no doubt they often migrate in advance of them.

The case of the young of the Cuckoo and other birds, and the fact that birds travel during the hours of darkness, convince me that migratory birds are endowed with "inherited but unconscious experience" (the phrase is Professor Newton's). This special sense of direction guides them, though they know it not, and enables them to reach the seasonal homes of their respective forebears, wherever these may be. I was convinced of this unconscious guidance by my singular experiences on board the lightship, to which allusion will be made anon.

Further important evidence in support of the conclusion that birds possess a special sense of direction, is afforded by the extensive migrations performed by the flightless Penguins. Sir J. C. Ross[1] informs us that he observed, in lat. 49° 17' S., long. 172° 28' W., penguins in great abundance going eastwards, which, he tells us, were no doubt proceeding to their breeding quarters, perhaps the Nimrod Islands. "It is a wonderful instinct," he remarks, "far beyond the powers of un-

[1] *A Voyage of Discovery and Research in the Southern and Antarctic Regions, during the years* 1839-1843, vol. ii., p. 138.

tutored reason, that enables these creatures to find their way, chiefly under water, several hundred miles, to their place of usual resort, as each succeeding spring season of the year arrives." On another occasion, Ross observed two penguins when more than a thousand miles from the nearest land (*loc. cit.*, vol. i. p. 227-8). The erratic wanderings of migratory birds, resulting in their appearance in countries far removed from their accustomed haunts, and off the routes followed to reach them, are in many cases to be attributed to their failure, from some cause or other, to inherit unimpaired this all-important faculty of unconscious orientation. The incentive to migrate, it must be admitted, is strong within them, or they would never occur in places so remote from the domains of their respective species.

Such facts as these most effectually dispose of the contentions advanced in favour of sight, individual memory, knowledge of landmarks, the aid derived from flying at a great height, and of experience imparted to the young by parental guidance. All these would be of no avail to a party of Penguins seeking their accustomed summer quarters when surrounded on all sides by hundreds of miles of the pathless, featureless ocean. They compel us to fall back upon a special sense of direction unconsciously exercised : no other explanation seems possible.

Some very remarkable experiments have recently been carried out by American ornithologists, which also go far to prove that birds are endowed with a "mysterious sense of direction." The following is quoted from Mr Chapman's article in *Bird Lore* (1908, p. 134) on experiments carried out by the Department of Marine Biology of the Carnegie Institution. From

these it appears that fifteen marked Noddy and Sooty
Terns were taken from their nesting haunts at Bird
Key, Tortugas, and released at distances varying from
20 to 850 statute miles, and that thirteen of them
returned to the Key. Among the thirteen were several
birds which had been taken by steamer as far north as
Cape Hatteras, before being freed. "This experiment,"
Mr Chapman proceeds to say, "is by far the most
important in its bearing on bird-migration of any with
which we are familiar. It was made under ideal conditions; neither the Noddy nor the Sooty Tern range,
as a rule, north of the Florida Keys. There is no
probability, therefore, that the individuals released had
ever been over the route before, and, for the same reason,
they could not have availed themselves of the experience
or example of migratory individuals of their own species ;
nor, since the birds were doubtless released in June or
July, was there any marked southward movement in the
line of which they might follow. Even had there been such
a movement, it is not probable that it would have taken
the birds south-west to the Florida Keys, and thence
west to the Tortugas. This marked change of direction,
occasioned by the water course, which the birds' feeding
habits forced them to take, removes the direction of the
wind as a guiding agency, while the absence of landmarks
over the greater portion of the journey makes it improbable that sight was of service in finding the way. . . .
We cannot but feel that the experiments with these birds
constitute the strongest argument for the existence of a
sense of direction as yet derived from the study of birds.
With this established, the so-called mystery becomes no
more a mystery than any other instinctive functional
activity."

CHAPTER III

THE BRITISH ISLES AND THEIR MIGRATORY BIRDS

IT may be asserted without fear of contradiction, that no country in the world is more favourably situated than our own for witnessing the movements of migratory birds; that there is none in which the many phases of the phenomenon are of a more varied nature; and none in which the subject has received greater attention.

The reason for this pre-eminence is not difficult to discover. The geographical position of our isles is favourable for the visits of migratory birds to a remarkable degree. They are interposed between the great south-western extension of Europe (the Iberian peninsula) and Scandinavia on the one hand, and the main mass of Continental Europe and the great islands of Iceland and Greenland on the other. Thus they are situated between the winter retreats and the summer quarters of hosts of birds, and lie in the direct course of vast multitudes of migrants which for weeks during each season of migration rush—northwards in spring and southwards in autumn—along our shores. Indeed, so central is our position, that it may be likened unto a great junction—one where many travellers meet and change their course to reach accustomed haunts in widely different lands beyond our shores.

In addition to affording a half-way house, as it were, for this wonderful concourse of transitory visitors, the British Isles have a great bird-population of their own, and the majority of their feathered citizens are migratory. This again is due to their central geographical position, which renders them eminently suitable both as a summer residence and a winter home.

Our remarkable climate, too, plays an important part. Its mild winter temperatures approaching, as they do, those prevailing in southern Europe, induce a number of species from the north to pass the winter with us which would not remain if our climatic conditions at that season were of the continental type in similar latitudes. MacKinder (*Britain and the British Isles*, p. 170) tells us that we are "placed not far from the centre of the area of abnormal warmth in winter, and the British winters are, in consequence, milder than those of any other region under the same northern latitudes."

Drayton in his *Polyolbion* (1613) thus quaintly sums up these peculiarities :—

> "Of Albion's glorious isle, the wonders whilst I write,
> The sundry varying soyles, the pleasures infinite ;
> Where heat kills not the cold, nor cold expels the heat,
> No calmes too mildly small, nor winds too roughly great ;
> Nor night doth hinder day, nor day the night doth wrong,
> The summer not too short, the winter not too long."

Further evidence of the singular wealth of migratory bird-life in the British Isles is afforded by the fact that more than one-half of the species which form the ornis of Europe are either regular or not infrequent visitors to our shores.

Small wonder, then, that with so many and varied migratory birds around them, the naturalists of our islands have found the subject of their movements such

an attraction, and that since the days of Gilbert White
so much attention has been bestowed on this singularly
fascinating branch of ornithology.

Now let us consider the various classes of migratory
birds to be observed in our islands, which include representatives of all the known phases of the migratory habit.

SUMMER VISITORS.[1]—In spring we welcome the
appearance of a number of birds, many of them of
delicate and graceful form and not a few of delightful
song, which arrive to spend the summer with us.
These visitors add immeasurably to the joyousness of
spring, and play an important part in its pageant. They
come to us from the south, in whose genial climes they
have passed the months of the drear northern winter,
having travelled far, some of them very far, to reach
our isles : not a few have winged their way from
Southern Africa ; many from the equatorial regions of
the same continent ; and many, again, from the countries
bordering upon the western shores of the Mediterranean
Sea. These birds rear their broods in our midst, and
during the autumn take their departure (as do their
offspring) to seek again their far-off winter retreats.
These summer visitors number over fifty different
species. They are a varied set of guests, and frequent
all manner of haunts. The Ring Ouzel seeks the
moorlands, the Wheatear the waste lands and downs,
the Willow Warbler, Blackcap, and Nightingale the
woodlands and copses, the Sedge Warbler the river
and brookside, the Swallow, Martin, and Swift the
haunts of man, the Flycatcher our gardens, the Whitethroat the hedgerows, the Nightjar the heath, the

[1] For a list of the British Summer Visitors, with an indication of their
winter quarters, see page 46.

Cuckoo almost everywhere, the Corn-Crake the grass-lands, the Sandpiper the river and the loch, the Dotterel the summits of fells and mountains, the Terns the sea shore and the off-lying islets. Some are widely distributed, others more or less local. Such are a few typical summer birds.

It is a noteworthy fact that all these summer visitors to our islands, and also the birds of passage from the north which traverse our shores in the autumn, have winter retreats in countries to the south of the British area, either in Southern Europe, or in Northern, Tropical, and Southern Africa.

None of the birds which breed in the British Isles, or indeed in the Northern Hemisphere, it should be remarked, nest during their sojourn in their southern winter quarters, though some of them were once thought to do so.

PARTIAL MIGRANTS.[1]—In addition to the summer visitors, the members of which depart from the British Islands at their wonted times, there are other birds usually regarded as resident species (such as the familiar Skylark, Song Thrush, Pied-Wagtail, Meadow Pipit, Lapwing, and others), but among which many individuals are migratory—especially those inhabiting the more northern and elevated districts. In reality these are SUMMER VISITORS to the British Isles, as are all, or nearly all, the members of the same species in corresponding latitudes on the Continent. These partial migrants, as they are termed, do not, as a rule, travel far beyond the British area to winter; and are the earliest among our migratory birds to leave us in the autumn and to return to their native haunts in spring.

It may be here remarked that in some cases where

[1] For list of British Partial Migrants, see page 49.

certain birds are observed throughout the year they are not necessarily represented in summer and winter by the same individuals, for the place of those which have passed the summer here is often taken in winter by immigrants from other areas, mainly from continental Europe.

The ordinary observer, until lately at least, never thought birds usually considered residents were represented in our islands by a sedentary and a migratory race.

LOCAL BRITISH MOVEMENTS.—In addition to the important migrations between the British Isles and the more or less far-off lands to the north, south, and east of them, much migration takes place in spring, autumn, and winter between various portions of the British area. These local movements are undertaken in spring to reach nesting haunts from winter quarters near at hand or somewhat far removed. In autumn the summer homes are quitted and a return made to winter retreats. In some cases these movements are from Great Britain to Ireland—the Sister Isle being the winter home. There are also smaller local winter movements, but these will be alluded to in another section. Thus there are innumerable minor migrations between all the parts of the British area, and hosts of Sea-Fowl on quitting the rocky fastnesses, in which the summer has been passed, spend the winter in the neighbouring seas.

WINTER VISITORS.[1]—In the autumn a great change takes place in the bird-life of our islands. The summer guests gradually slip away, and another set of birds, also seeking winter quarters, arrives from Northern and

[1] For a list of the British Winter Visitors, with an indication of their summer haunts, see page 50.

Central Europe, Iceland, Greenland, and other islands in the Arctic Ocean. These hardy northern races find the climatic conditions of the British Islands sufficiently genial to afford cold-weather retreats, and remain with us through the winter months. On the return of spring they leave us, to repair to their native haunts to nest. These winter visitors number over one hundred species, and include such northern forms as the Fieldfare, Brambling, Short-eared Owl, various species of Duck, Geese, and Swans, some Gulls, and many Waders; while from Central Europe, as well as from the north, come numerous Rooks, Starlings, Tree Sparrows, Skylarks, Lapwings, etc.

BIRDS OF PASSAGE.[1]—Though they are far and away the most numerous class of migrants that visit our shores, the birds of passage are less familiar than either the summer or the winter visitors. This is due to the fact that they do not remain long with us, and are chiefly confined to the coast and its vicinity during their sojourn, and consequently they do not come under the notice of those who are not specially interested in bird-movements. They are birds which have their nesting haunts chiefly in the vast regions lying to the north of the British area, and their winter quarters to the south of us; while a number are natives of Central Europe. They appear on and traverse our shores in spring, when en route for their summer homes, and again in the autumn, when returning to winter retreats. During these visits they do not, as a rule, tarry long with us, especially in the spring, but make our shores a sort of half-way house, where they rest and refresh themselves

[1] For a list of the British Birds of Passage, with an indication of their summer and winter quarters, see page 55.

ere they resume what are in many cases their long and intricate journeys. In the ranks of these birds of passage are to be found the greatest of all bird-travellers, some of which—as the Arctic Tern—all but extend their journeys from pole to pole; others from Siberia to the utmost limits of Africa—as the Curlew Sandpiper; others again, from Northern Europe to Equatorial and Southern Africa—as the Swift, etc. The Curlew Sandpiper is, perhaps, the greatest of all feathered voyagers. This species has its summer haunts in western Siberia, where it nests on the tundras fringing the Arctic Ocean; yet its winter range extends to Cape Colony, Madagascar, Patagonia, Tasmania, and the Malay Archipelago. To reach these far-off cold-weather retreats, it crosses the lofty Himalayas; traverses the course of the great rivers of Northern Asia, and of the Volga, Rhone, and Nile, and skirts the coasts of Norway, Britain, Western Europe and Africa, and China. Thus during each year certain Curlew Sandpipers perform journeys equal to a voyage round the world! Not all travel thus far, for this bird has winter haunts on both sides of the Mediterranean, in Central Africa, on the coasts of India, and in the Philippines. Indeed, in this and not a few other cases, it would seem as if these bird-travellers said (with Wordsworth):—

> Wings have we, and as far as we can go
> We may find pleasure.

The fact that this and other migratory birds cover such extensive areas during the winter indicates that they, and perhaps all migrants, have a much wider range at that season than in summer: they may be more nomadic in winter than at other times.

The birds of passage are in most cases the latest to

appear on our shores in spring and the last to depart from them in the autumn. That this should be so is readily explained by the fact that the summer homes of most of them lie in the northern lands, where the nesting period is naturally much later than in our own country. The White and Grey-headed Wagtails, Blue-throated Warbler, Honey Buzzard, Little Stint, and Curlew Sandpiper are examples of these transitory migrants.

Before dismissing the birds of passage, it must be made clear that in addition to the species which occur with us only as such, vast numbers of individuals belonging to almost all the species which are summer[1] and winter visitors to our islands occur also on passage on our shores. Thus many Swallows summer up to a high latitude in Scandinavia, and many Fieldfares winter in Southern Europe, and both these species traverse the British coastlines to get to their respective destinations.

The various migrations performed by these summer visitors, winter visitors, birds of passage, and partial migrants are phases of two seasonal movements only— all the birds are proceeding to summer haunts in spring, all to winter retreats in autumn.

In the Local Movements we see migration in its simplest aspect, in the Birds of Passage in the highest stage of its development.

WINTER MOVEMENTS.—In addition to the *regular migrations* above described, there are cold-weather movements due to, and dependent upon, the climatic

---

[1] The following summer visitors to the British Isles do *not* occur with us as birds of passage to and from the north :—Nightingale, Marsh Warbler, Stone-Curlew, and Kentish Plover. The Red-backed Shrike, Lesser Whitethroat, Wryneck, and Ruff, which occur on passage in Scotland, are not summer visitors to Northern Britain.

conditions of the season. Usually during each winter there are outbreaks of cold of a more or less severe nature and of longer or shorter duration. Such outbreaks compel certain resident species, and also some of our winter guests, to shift from the areas affected to others where more favourable conditions prevail, and where food, the main incentive for the movements, can be obtained. Some of these evicted birds find refuges in milder districts, if such there be, within the British Isles, especially in the western districts and Ireland, while others proceed to Southern Europe. During winters of exceptional severity, and particularly when the entire British area is affected and food therefore almost unobtainable, many birds which have endeavoured to brave the storm with us perish in great numbers, even those which have sought what are usually the mildest portions of our islands—namely, the south-western portions of England and Ireland. The species which are most affected by these cold snaps include, among others, the Mistle Thrush, Song Thrush, Redwing, Fieldfare, Blackbird, Greenfinch, Starling, Skylark, Water Rail, Lapwing, Curlew, Snipe, and Woodcock.

These winter movements do not occur in the continental areas to a like extent. There the climatic conditions of the season are more stable, and are not subject to the ups and downs to be found with us. Consequently such emigrations are more or less exceptional, and it is only in winters of remarkable severity that the birds accustomed to winter in Scandinavia and in Central Europe are compelled to quit their retreats. When this does occur there is a renewal on our shores of movements similar to the autumnal influx of winter visitors from the north-east and east, but on the part of very much smaller numbers both of species and of individuals.

CASUAL VISITORS.—Another class of bird-visitors to our islands now claims attention—namely, the Irregular or Casual visitors. Each year a number of Continental, Asiatic, African, and American species appear on our shores as stragglers, many of them after having performed remarkable peregrinations. Though rich in the number of forms, these waifs are comparatively poor in individuals, and mostly belong to species which annually perform extensive migratory journeys. In the great majority of cases they arrive in the autumn and are juveniles ; and their appearance among us is, probably, in most instances due to the errors or indiscretions of youth : they have failed to follow the right course leading to the usual winter retreats resorted to by their race. They inherited the migratory impulse in its undiminished force, but the unconscious experience has for some unknown reason been denied them, and they mostly share the fate which nature rigorously imposes on the "unfit." In some cases, possibly often, they may have been overtaken by adverse weather when making for their accustomed seasonal haunts, and thus have been driven out of their course, to be stranded in localities where they would not otherwise have occurred.

In spring much smaller numbers of these waifs appear on our shores. Their appearance at that season far beyond the bounds of their usual summer quarters may, in some instances, be accounted for, perhaps, by an excess of zeal having led them to overshoot, as it were, the ordinary limits of their range ; while in others, strong winds may, as in the case of some of the autumn stragglers, have carried them more or less out of their ordinary course when on passage.

These casuals form a very numerous section of the British avifauna, for no less than two hundred different kinds have from time to time occurred in our isles. The American species which find their way to our shores *unaided*, are birds which have a high northern summer range, and they doubtless reach us after having travelled by way of Greenland, Iceland, and the Faroes, not by an impossible passage across the open Atlantic. In this way their voyages are not more wonderful than those annually performed by the Wheatear, Redwing, Whimbrel, and others along similar lines of flight.

It is not necessary to give a complete list of the species forming the imposing array of Irregular Visitors. The following selection has been prepared with the view of affording some indication of the wide range of the orders to which these gypsy-migrants belong, and whence they have wandered to reach our isles.

The following are the more important—*i.e.*, most frequent—visitors from CONTINENTAL EUROPE, whence, naturally, most of these casuals come :—

NUTCRACKER, *Nucifraga caryocatactes.*
ROSE-COLOURED STARLING, *Pastor roseus.*
LITTLE BUNTING, *Emberiza pusilla.*
RUSTIC BUNTING, *Emberiza rustica.*
BLACK-HEADED BUNTING, *Emberiza melanocephala.*
PINE-GROSBEAK, *Pinicola enucleator.*
SERIN FINCH, *Serinus serinus.*
SHORT-TOED LARK, *Calandrella brachydactyla.*
CRESTED LARK, *Galerida cristata.*
RED-THROATED PIPIT, *Anthus cervinus.*
TAWNY PIPIT, *Anthus campestris.*
RICHARD'S PIPIT, *Anthus richardi.*
WATER-PIPIT, *Anthus spinoletta.*

LESSER GREY SHRIKE, *Lanius minor.*
WOODCHAT SHRIKE. *Lanius senator.*
GREAT REED-WARBLER, *Acrocephalus turdoides.*
MELODIOUS WARBLER, *Hypolais polyglotta.*
SIBERIAN CHIFF-CHAFF, *Phylloscopus tristis.*
ALPINE ACCENTOR, *Accentor collaris.*
WHITE-SPOTTED BLUETHROAT, *Cyanecula cyanecula.*
DESERT WHEATEAR, *Saxicola deserti.*
BLACK-BELLIED DIPPER, *Cinclus cinclus.*
RED-BREASTED FLYCATCHER, *Muscicapa parva.*
ALPINE SWIFT, *Cypselus melba.*
ROLLER, *Coracias garrulus.*
TENGMALM'S OWL, *Nyctala tengmalmi.*
SCOP'S OWL, *Scops scops.*
SPOTTED EAGLE, *Aquila maculata.*
ICELAND FALCON, *Falco islandus* (from Iceland).
HARLEQUIN DUCK, *Cosmonetta histrionica* (Iceland).
RED-CRESTED POCHARD, *Netta rufina.*
FERRUGINOUS DUCK, *Nyroca nyroca.*
RUDDY SHELD-DUCK, *Tadorna casarca.*
GLOSSY IBIS, *Plegadis falcinellus.*
BLACK STORK, *Ciconia nigra.*
PURPLE HERON, *Ardea purpurea.*
GREAT WHITE HERON, *Herodias alba.*
LITTLE EGRET, *Garzetta garzetta.*
SQUACCO HERON, *Ardeola ralloides.*
NIGHT-HERON, *Nycticorax nycticorax.*
LITTLE BITTERN, *Ardetta minuta.*
BAILLON'S CRAKE, *Porzana intermedia.*
LITTLE CRAKE, *Porzana parva.*
COMMON CRANE, *Grus grus.*
LITTLE BUSTARD, *Otis tetrax.*
PRATINCOLE, *Glareola pratincola.*
LITTLE RINGED PLOVER, *Ægialitis dubia.*
STILT, *Himantopus himantopus.*
BROAD-BILLED SANDPIPER, *Limicola platyrhyncha.*
WHISKERED TERN, *Hydrochelidon hybrida.*
WHITE-WINGED BLACK TERN, *H. leucoptera.*
GULL-BILLED TERN, *Sterna anglica.*
CASPIAN TERN, *S. caspia.*

The Asiatic Birds which have visited our isles include :—

YELLOW-BROWED BUNTING, *Emberiza chrysophrys*.
SIBERIAN MEADOW-BUNTING, *E. cioides*.
LARGE-BILLED REED-BUNTING, *E. palustris*.
SYKES' WAGTAIL, *Motacilla beema*.
PALLAS' GRASSHOPPER WARBLER, *Locustella certhiola*.
RADDE'S BUSH-WARBLER, *Lusciniola schwarzi*.
PALLAS' WILLOW-WARBLER, *Phylloscopus proregulus*.
DUSKY THRUSH, *Turdus dubius*.
BLACK-THROATED THRUSH, *Turdus atrigularis*.
WHITE'S THRUSH, *Geocichla varia*.
EASTERN PIED WHEATEAR, *Saxicola pleschanka*.
BROWN FLYCATCHER, *Muscicapa latirostris*.
SPINE-TAILED SWIFT, *Chætura caudacuta*.
RED-BREASTED GOOSE, *Branta ruficollis*.
PALLAS' SAND-GROUSE, *Syrrhaptes paradoxus*.
MACQUEEN'S BUSTARD, *Houbara macqueeni*.
SIBERIAN PECTORAL SANDPIPER, *Tringa acuminata*.

The majority of these species have appeared on one or two occasions only.

The essentially AFRICAN VISITORS are very few, but include :—

MASKED SHRIKE, *Lanius nubicus*.
CREAM-COLOURED COURSER, *Cursorius gallicus*.

NORTH AMERICA has contributed numerous species, most of which are extremely irregular in their visits, while a few, such as the Pectoral Sandpiper, are of almost annual occurrence. They are :—

AMERICAN PIPIT, *Anthus pennsylvanicus*.
YELLOW-BILLED CUCKOO, *Coccyzus americanus*.
BLACK-BILLED CUCKOO, *C. erythrophthalmus*.
HAWK-OWL, *Surnia caparoch*.
SNOW-GOOSE, *Chen hyperboreus*.
GREATER SNOW-GOOSE, *C. nivalis*.
AMERICAN WIGEON, *Mareca americana*.

AMERICAN TEAL, *Nettion carolinense.*
BLUE-WINGED TEAL, *Querquedula discors.*
BUFFLE-HEADED DUCK, *Clangula albeola.*
HOODED MERGANSER, *Mergus cucullatus.*
SURF-SCOTER, *Œdemia perspicillata.*
AMERICAN BITTERN, *Botaurus lentiginosus.*
CAROLINA RAIL, *Porzana carolina.*
AMERICAN GOLDEN PLOVER, *Charadrius dominicus.*
KILDEER PLOVER, *Ægialitis vocifera.*
PECTORAL SANDPIPER, *Tringa maculata.*
BONAPARTE'S SANDPIPER, *T. fuscicollis.*
BAIRD'S SANDPIPER, *T. bairdi.*
AMERICAN STINT, *T. minutilla.*
SEMIPALMATED SANDPIPER, *Ereunetes pusillus.*
BARTRAM'S SANDPIPER, *Bartramia longicauda.*
BUFF-BREASTED SANDPIPER, *Tringites rufescens.*
SPOTTED SANDPIPER, *Totanus macularius.*
SOLITARY SANDPIPER, *T. solitarius.*
YELLOWSHANKS, *T. melanoleucus* and *T. flavipes.*
RED-BREASTED SANDPIPER, *Macrorhamphus griseus.*
ESKIMO CURLEW, *Numenius borealis.*
BONAPARTE'S GULL, *Larus philadelphia.*

From the ARCTIC REGIONS we have received :—

GREENLAND REDPOLL, *Acanthis hornemanni.*
GREENLAND FALCON, *Falco candicans.*
SNOWY OWL, *Nyctea nyctea.*
ROSS' GULL, *Rhodostethia rosea.*
SABINE'S GULL, *Xema sabini.*
KING-EIDER, *Somateria spectabilis.*
STELLER'S EIDER, *Somateria stelleri.*
ADAM'S DIVER, *Colymbus adamsi.*
BRÜNNICH'S GULLEMOT, *Uria lomvia.*

Among the OCEAN WANDERERS which have appeared
on our shores are :—

BLACK-BROWED ALBATROS, *Diomedea melanophrys.*
MADEIRAN STORM-PETREL, *Oceanodroma castro.*
WILSON'S PETREL, *Oceanites oceanicus.*
FRIGATE-PETREL, *Pelagodroma marina.*

LEVANTINE SHEARWATER, *Puffinus yelkouanus.*
LITTLE DUSKY SHEARWATER, *P. bailloni.*
CAPPED PETREL, *Œstrelata hæsitata.*
COLLARED PETREL, *Œ. brevipes.*
SCHLEGEL'S PETREL, *Œ. neglecta.*
BULWER'S PETREL, *Bulweria bulweri.*

IRELAND.—All these phases of migration are observed in the Sister Isle. Ireland, too, has a great bird-population, of which a considerable section is migratory, and much migration is witnessed on her shores. She is not favoured to the same extent as Great Britain so far as the numbers of her Summer Visitors are concerned ; nor do her shores lie as directly in the course of so many Birds of Passage ; nor, again, thanks to her milder climate, are there so many migratory individuals (Partial Migrants) to be found in the ranks of her resident species. On the other hand, in addition to a considerable number of Winter Visitors from Northern Europe she receives in autumn, thanks again to her more genial climate, a great number of winter guests from less favoured portions of the British Area—such as Song Thrushes, Blackbirds, Chaffinches, Starlings, Skylarks, Lapwings and others ; while in seasons of severe cold or much snow, her hospitable shores, especially those of the west, afford the safest retreats in our area.

The movements of these several groups of migrants may be thus summarised under their seasons :—

SPRING.

1. Local movements from British winter retreats to British summer haunts.
2. The return from their continental winter quarters of the Partial Migrants.
3. The arrival in our islands from their southern winter retreats of Summer Visitors.

4. The departure from our islands of the Winter Visitors for their summer haunts.
5. The appearance on and journeys along our shores of Birds of Passage from the south, bound for summer haunts beyond our isles.
6. The return of the birds evicted by the severe weather of the previous winter.
7. The appearances of Casual Visitors.

AUTUMN.

1. Local movements from British summer haunts to British winter retreats.
2. The departure from our islands of the Summer Visitors for their winter retreats.
3. The departure for the winter of the Partial Migrants.
4. The arrival from the north and east of Winter Visitors.
5. The appearance on our shores of Birds of Passage en route from northern and eastern summer haunts to southern winter quarters.
6. The appearances of Casual Visitors.

WINTER.

1. The emigration of would-be Resident and Winter Visitors, through pressure of severe climatic conditions.
2. The arrival from the Continent of Immigrants similarly evicted by the weather.
3. The appearances of Casual Visitors.

Most of the species play many of these rôles as British migrants, as will be learned from the histories of the movements of the several species which will be treated of as typical migrants. Thus, for example, the Song Thrush figures as a summer visitor, a partial migrant, a bird of passage, a winter visitor, and a winter emigrant through eviction. Needless to remark, it is also a permanent resident in our islands, even in Scotland.

Not only do we witness much migration in the British Islands of a very varied nature, but often through a combination of influences, in which meteorological conditions play an important part, more than one movement, as we shall see, may be in progress simultaneously, —a circumstance which adds much to the already complicated series of phenomena, and to the bewilderment of the observer.

APPENDIX I.—LIST OF SUMMER VISITORS

The following is an enumeration of the various species of birds which are Summer Visitors to the British Islands. Many of them are very widely diffused over our area, such as the Swallow, while on the other hand a few are extremely circumscribed in their distribution, and of these the Marsh Warbler affords a good example.

An indication of the general WINTER RETREATS in the western regions of the Old World is given for each species. In some portion of these the cold season is passed; but at present we have no definite knowledge where the British members of any species spend the winter. It is highly probable, however, that as each has its particular summer home, so has it also a predilection for definite winter haunts :—

GOLDEN ORIOLE, *Oriolus oriolus.*—Tropical and South Africa.
WHITE WAGTAIL, *Motacilla alba.*—Southern Europe, North and North-tropical Africa.
YELLOW WAGTAIL, *Motacilla rayi.*—North-tropical, Equatorial, and Southern Africa.
BLUE-HEADED WAGTAIL, *Motacilla flava.*—Tropical, Equatorial, and South Africa.
TREE-PIPIT, *Anthus trivialis.*—South Europe, Northern and Tropical Africa to Transvaal.
RED-BACKED SHRIKE, *Lanius collurio.*—Tropical, Equatorial, and South Africa to Cape Colony.

WHITETHROAT, *Sylvia sylvia.*—Tropical and Equatorial Africa to Damara Land.

LESSER WHITETHROAT, *Sylvia curruca.*—Northern and North-tropical Africa.

BLACKCAP, *Sylvia· atricapilla.* — Southern Europe, Northern and Tropical Africa.

GARDEN-WARBLER, *Sylvia borin.*—Tropical, Equatorial, and South Africa to Damara Land and Natal.

WOOD-WARBLER, *Phylloscopus sibilatrix.*—North-tropical and Equatorial Africa to Abyssinia and Congo, Madeira and Canaries.

WILLOW-WARBLER, *Phylloscopus trochilus.*—North-western tropical, Equatorial, and South Africa to Cape Colony.

CHIFFCHAFF, *Phylloscopus collybita.*—Southern Europe, Northern and North-tropical Africa (Sudan).

SEDGE-WARBLER, *Acrocephalus schœnobœnus.*—Tropical and Equatorial Africa, as far south as Damara Land and Transvaal.

REED-WARBLER, *Acrocephalus streperus.*—South Europe, Northern Africa, Equatorial Africa to Gambia.

MARSH-WARBLER, *Acrocephalus palustris.*—Tropical and Equatorial Africa, Southern Africa.

GRASSHOPPER-WARBLER, *Locustella nœvia.*—South-western Europe, North-western Africa.

RING - OUZEL, *Turdus torquatus.* — Southern Europe, Northern Africa.

NIGHTINGALE, *Luscinia megarhynchus.*—Tropical Africa.

REDSTART, *Ruticilla phœnicurus.*—Oases of the Sahara, Northern Tropical Africa.

WHEATEAR, *Saxicola œnanthe.* — Tropical and Equatorial Africa (Senegal, Gambia, Sudan to Zambesi).

WHINCHAT, *Pratincola rubetra.*—Oases of the Sahara, North-tropical Africa.

SPOTTED FLYCATCHER, *Muscicapa grisola.*—South - tropical and Southern Africa.

PIED FLYCATCHER, *Muscicapa atricapilla.*—Northern Tropical and Equatorial Africa.

SWALLOW, *Hirundo rustica.*—Tropical, Central, and South Africa.

HOUSE-MARTIN, *Chelidon urbica.*—Equatorial, South-tropical, and Southern Africa.

SAND-MARTIN, *Cotile riparia.* — North- and South - tropical, and Equatorial Africa.

48 STUDIES IN BIRD-MIGRATION

WRYNECK, *Iÿnx torquilla.*—Northern and North-tropical Africa.

SWIFT, *Cypselus apus.*—North- and South-tropical, Equatorial, and South Africa.

NIGHTJAR, *Caprimulgus europæus.*—Equatorial, South-tropical, and Southern Africa.

CUCKOO, *Cuculus canorus.*—Equatorial, South-tropical, and Southern Africa.

MONTAGU'S HARRIER, *Circus cineraceus.*—Tropical, Equatorial, and South Africa to Cape Colony.

HOBBY, *Falco subbuteo.*—Northern, Tropical, and Equatorial Africa to Rhodesia and Damara Land.

HONEY-BUZZARD, *Pernis apivorus.*—Tropical and Southern Africa.

OSPREY, *Pandion haliaëtus.*—Southern Europe; Northern, Tropical, Equatorial, and South Africa.

GARGANEY, *Querquedula circia.*—Southern Europe, North and North-tropical Africa.

TURTLE-DOVE, *Turtur turtur.*—Equatorial Africa.

QUAIL, *Coturnix coturnix.*—Southern Europe, Northern and Tropical Africa.

CORN-CRAKE, *Crex crex.*—Tropical and South Africa to Cape Colony.

STONE-CURLEW, *Œdicnemus œdicnemus.*—Southern Europe and Northern Africa.

DOTTEREL, *Eudromias morinellus.*—Southern Europe and Northern Africa.

KENTISH PLOVER, *Ægialitis cantiana.*—Southern Europe; Northern, Central, and South-tropical Africa.

RED-NECKED PHALAROPE, *Phalaropus hyperboreus.*—Seas of South-western Europe.

COMMON SANDPIPER, *Totanus hypoleucus.*—Southern Europe, Northern, Tropical, Equatorial, and Southern Africa.

WHIMBREL, *Numenius phæopus.*—Southern Europe, Africa to Cape Colony.

SANDWICH TERN, *Sterna cantiaca.*—Mediterranean, West African Seas to Cape and Natal.

COMMON TERN, *Sterna fluviatilis.*—Tropical Atlantic to Seas of Cape Colony.

ARCTIC TERN, *Sterna macrura.*—Tropical Atlantic and Antarctic Ocean (Weddell Sea).

ROSEATE TERN, *Sterna dougalli.*—Tropical African Seas.

LITTLE TERN, *Sterna minuta.*—Seas of West-tropical Africa.

APPENDIX II.—LIST OF PARTIAL MIGRANTS

The following are also SUMMER VISITORS to the British Isles, being the migratory representatives of species which are residents in our islands at all seasons. The precise Winter Retreats of these Partial Migrants have not been ascertained, but where known are in countries not far to the south of the British area.[1] Some of these belong to races which are peculiar to the British Isles and are indicated by trinomial scientific names in the list.

ROOK, *Corvus frugilegus.*
STARLING, *Sturnus vulgaris.*
GOLDFINCH, *Carduelis carduelis britannicus.*
GREENFINCH, *Chloris chloris.*
LINNET, *Acanthis cannabina.*
TWITE, *Acanthis flavirostris.*
SKYLARK, *Alauda arvensis.*
PIED WAGTAIL, *Motacilla lugubris.*
GREY WAGTAIL, *Motacilla boarula.*
MEADOW-PIPIT, *Anthus pratensis.*
GOLDCREST, *Regulus regulus anglorum.*
SONG-THRUSH, *Turdus musicus clarkei.*
MISTLE-THRUSH, *Turdus viscivorus.*
BLACKBIRD, *Turdus merula.*
REDBREAST, *Erithacus rubecula melophilus.*
STONECHAT, *Pratincola rubicola hibernans.*
HEDGE-ACCENTOR, *Accentor modularis occidentalis.*
KESTREL, *Falco tinnunculus.*
HERON, *Ardea cinerea.*
MALLARD, *Anas boscas.*
TEAL, *Nettion crecca.*

[1] Starlings marked in Britain have been recovered in France, Greenfinch in France, Linnet in France, Pied Wagtail in Portugal, Meadow-Pipit in Portugal, Song-Thrushes in France and Portugal, Mallard and Teal in Germany, Wigeon in Holland, Lapwings in France and Portugal, Woodcock in Portugal, Black-headed Gull in France, and Lesser Black-backed Gulls on the coasts of France and Portugal.

WIGEON, *Mareca penelope.*
RINGED PLOVER, *Ægialitis hiaticola.*
GOLDEN PLOVER, *Charadrius pluvialis.*
LAPWING, *Vanellus vanellus.*
OYSTER-CATCHER, *Hæmatopus ostralegus.*
WOODCOCK, *Scolopax rusticula.*
COMMON SNIPE, *Gallinago gallinago.*
DUNLIN, *Tringa alpina.*
REDSHANK, *Totanus calidris.*
CURLEW, *Numenius arquata.*
BLACK-HEADED GULL, *Larus ridibundus.*
LESSER BLACK-BACKED GULL, *Larus fuscus.*

Probably a number of other species which are always with us have also a migratory race, the members of which leave us in the autumn and return to our islands in spring.

APPENDIX III.—LIST OF WINTER VISITORS

The following are Winter Visitors to the British Islands, with an indication of their SUMMER HAUNTS in the western regions of the Old World, whence they may have come to pass the cold season with us.[1] Those marked * are also Resident species in the British Isles.

*ROOK, *Corvus frugilegus.*—Scandinavia, Central Europe.
*CARRION-CROW, *Corvus corone.*—Central Europe.
*GREY CROW, *Corvus cornix.*—Scandinavia, Russia, Central Europe.
*JACKDAW, *Corvus monedula.*—Northern Continental and Central Europe.
*STARLING, *Sturnus vulgaris.*—Norway, Central Europe.
*CHAFFINCH, *Fringilla cœlebs.*—Northern Continental Europe.
BRAMBLING, *Fringilla montifringilla.*—Norway, Lapland, Northern Russia.

[1] A number of the species also spend the summer in Siberia. These Eastern representatives usually seek winter retreats in Southern Asia, the Malay Islands, and even in Australia. Fewer have representatives also in North America, which move southwards to winter quarters in the New World. Some of the Western Siberian birds may find their way to our islands.

*SISKIN, *Spinus spinus.*—Norway, Mid Sweden, Russia.

MEALY REDPOLL, *Acanthis linaria.*—Iceland, Scandinavia, North Russia, Finland, Baltic Provinces (also North Asia and America).

*TWITE, *Acanthis flavirostris.*—Scandinavia.

*TREE-SPARROW, *Passer montanus.*—Scandinavia, Russia, Central Europe.

*GREENFINCH, *Chloris chloris.*—Scandinavia, North Russia.

CONTINENTAL CROSSBILL, *Loxia curvirostra.*—Scandinavia and Russia, where it is also resident.

*YELLOW BUNTING, *Emberiza citrinella.*—Northern Continental Europe.

*REED-BUNTING, *Emberiza schœniclus.*—Northern Continental Europe.

LAPLAND BUNTING, *Calcarius lapponicus.*—Northern Europe, Greenland, Kolguev, Novaya Zemlya (also North Asia and America).

*SNOW-BUNTING, *Plectrophenax nivalis.* — Arctic and Sub-Arctic Continental Europe, Greenland, Iceland, Faroes, Spitzbergen, Franz Josef Land, Novaya Zemlya, etc. (also Arctic Asia and America).

*SKYLARK, *Alauda arvensis.*—Northern Continental and Central Europe.

SHORE-LARK, *Otocorys alpestris.*—Northern Europe, including Kolguev, Novaya Zemlya, Siberia.

*MEADOW-PIPIT, *Anthus pratensis.*—Iceland, Northern Continental Europe.

*ROCK-PIPIT, *Anthus obscurus.*—Western Scandinavia to the White Sea.

CONTINENTAL GOLDCREST, *Regulus regulus.*—Norway, Sweden, Finmark, Lapland, North Russia.

GREAT GREY SHRIKE, *Lanius excubitor.*—Scandinavia, North Russia; elsewhere more or less resident.

WAXWING, *Ampelis garrulus.*—North Finland and North Russia (also Siberia and Arctic America).

*MISTLE-THRUSH, *Turdus viscivorus.*—Northern Continental and Central Europe.

CONTINENTAL SONG-THRUSH, *Turdus musicus.*—Northern and Temperate Continental Europe and Siberia (Scandinavia to Lake Baikal).

REDWING, *Turdus iliacus.*—Iceland, Northern Continental Europe and West Siberia (Norway to the Yenesei).

FIELDFARE, *Turdus pilaris.*—Northern Continental Europe and West Siberia (Norway to the Yenesei).

*BLACKBIRD, *Turdus merula.*—Scandinavia to 67° N., Central Russia.

CONTINENTAL REDBREAST, *Erithacus rubecula.*—Northern Continental
Europe.

BLACK REDSTART, *Ruticilla titys.*—Central Europe, resident in
Southern Europe.

CONTINENTAL HEDGE-ACCENTOR, *Accentor modularis.*—Scandinavia,
Russia to 60° N.

*WREN, *Troglodytes troglodytes.*—Northern Continental Europe.

CONTINENTAL GREAT SPOTTED WOODPECKER, *Dendrocopus major.*—
Scandinavia to 70° N.; Russia to 64° N.

*LONG-EARED OWL, *Asio otus.*—Scandinavia and North Russia.

*SHORT-EARED Owl, *Asio accipitrinus.*—Scandinavia, Russia, etc.
(Siberia.)

ROUGH-LEGGED BUZZARD, *Archibuteo lagopus.*—Scandinavia, Northern
Russia, Waigatsch. (Siberia.)

*PEREGRINE FALCON, *Falco peregrinus.*—Northern Continental Europe.

*MERLIN, *Falco æsalon.*—Iceland, Northern Continental Europe.

*KESTREL, *Falco tinnunculus.*—Scandinavia.

*HERON, *Ardea cinerea.*—Scandinavia.

BITTERN, *Botaurus stellaris.*—Central Europe, Southern Sweden.

*GREY LAG-GOOSE, *Anser anser.*—Iceland, Scandinavia, Russia, etc.

WHITE-FRONTED GOOSE, *Anser albifrons.*—Greenland, Iceland,
Kolguev, Novaya Zemlya. (Siberia.)

BEAN-GOOSE, *Anser segetum.*—Greenland, Northern Scandinavia,
Russia, Kolguev, Novaya Zemlya. (Siberia.)

PINK-FOOTED GOOSE, *Anser brachyrhynchus.*—Iceland, Spitzbergen.

BERNACLE-GOOSE, *Branta leucopsis.*—Greenland, Spitzbergen.

BRENT GOOSE, *Branta bernicla.*—Greenland, Spitzbergen, Franz Josef
Land, Kolguev. (Siberia.)

WHOOPER, *Cygnus musicus.*—Iceland, Northern Scandinavia and
Russia. (Siberia.)

BEWICK'S SWAN, *Cygnus bewicki.*—North-east Russia, Kolguev,
Novaya Zemlya, Waigatsch. (Siberia.)

*SHELD-DUCK, *Tadorna tadorna.*—Scandinavia.

*MALLARD, *Anas boscas.*—Southern Greenland, Iceland, Northern
Europe.

*GADWALL, *Anas strepera.*—Iceland, North Russia, Central Europe.

*SHOVELER, *Spatula clypeata.*—Northern and Central Europe.

*PINTAIL, *Dafila acuta.*—Iceland, Faroes, Northern and Central
Europe.

*TEAL, *Nettion crecca.*—Iceland, Northern Continental Europe.

*WIGEON, *Mareca penelope.*—Iceland, Kolguev, Waigatsch, Northern Continental Europe.

*POCHARD, *Fuligula ferina.*—Central Europe.

*TUFTED DUCK, *Fuligula fuligula.*—Scandinavia, Russia.

*SCAUP-DUCK, *Fuligula marila.*—Iceland, Faroes, Northern Europe. (Siberia.)

GOLDENEYE, *Clangula glaucion.*—Iceland, Northern Scandinavia, Russia, and Central Europe.

LONG-TAILED DUCK, *Harelda glacialis.*—Greenland, Iceland, Faroes, Spitzbergen, Novaya Zemlya, Northern Continental Europe. (Siberia.)

*EIDER-DUCK, *Somateria mollissima.*—Greenland, Iceland, Faroes, Spitzbergen, Franz Josef Land, Novaya Zemlya, Coasts of Northern Europe.

*COMMON SCOTER, *Œdemia nigra.*—Iceland, Kolguev, Novaya Zemlya, Waigatsch, Northern Continental Europe. (Western Siberia.)

VELVET-SCOTER, *Œdemia fusca.*—Northern Continental Europe. (Western Siberia.)

*GOOSANDER, *Mergus merganser.*—Iceland, Novaya Zemlya, Northern Continental Europe. (Siberia.)

*RED-BREASTED MERGANSER, *Mergus serrator.*—Southern Greenland, Iceland, Faroes, Northern Continental Europe. (Siberia.)

SMEW, *Mergus albellus.*—Lapland, Northern Russia. (Siberia.)

*RING-DOVE, *Columba palumbus.*—Northern Continental Europe.

*WATER-RAIL, *Rallus aquaticus.*—Iceland, Southern and Central Scandinavia.

*COOT, *Fulica atra.*—Southern Scandinavia.

*GOLDEN PLOVER, *Charadrius pluvialis.*—Iceland, Faroes, Northern Continental Europe.

GREY PLOVER, *Squatarola helvetica.*—North-east Russia, Kolguev, Novaya Zemlya. (Siberia.)

*LAPWING, *Vanellus vanellus.*—Northern Continental and Central Europe.

*RINGED PLOVER, *Ægialitis hiaticola.*—Greenland, Iceland, Spitzbergen, Novaya Zemlya, Kolguev, Northern Europe.

TURNSTONE, *Strepsilas interpres.*—Greenland, Iceland, Kolguev, Novaya Zemlya, Coasts of Norway, Northern Europe, and Asia.

*OYSTER-CATCHER, *Hæmatopus ostralegus.* — North-western Europe, Iceland.

I.   D 2

54   STUDIES IN BIRD-MIGRATION

Grey Phalarope, *Phalaropus fulicarius.*—Greenland, Iceland, Spitzbergen, Novaya Zemlya. (Siberia.)

*Woodcock, *Scolopax rusticula.*—Northern Continental Europe.

Jack Snipe, *Gallinago gallinula.*—Northern Scandinavia, Northwestern Russia. (Siberia.)

*Common Snipe, *Gallinago gallinago.*—Iceland, Faroes, Northern Continental Europe.

*Dunlin, *Tringa alpina.*—Eastern Greenland, Iceland, Faroes, Kolguev, Waigatsch, Northern Continental Europe.

Purple Sandpiper, *Tringa maritima.*—Greenland, Iceland, Faroes, Spitzbergen, Franz Josef Land, Novaya Zemlya, Coasts of Northern Europe and Asia.

Knot, *Tringa canutus.*—Arctic America, Greenland, Tundras of Siberia.

Sanderling, *Calidris arenaria.*—Greenland, Spitzbergen, Tundras of Siberia.

*Redshank, *Totanus calidris.*—Iceland, Northern Continental Europe.

Bar-tailed Godwit, *Limosa lapponica.*—Russian Lapland, Western Siberia.

*Curlew, *Numenius arquata.*—Northern Continental and Central Europe.

*Common Gull, *Larus canus.*—Scandinavia, Northern Russia.

*Herring-Gull, *Larus argentatus.*—Coast of Scandinavia.

*Greater Black-backed Gull, *Larus marinus.*—Iceland, Scandinavia, North Russia.

Glaucous Gull, *Larus glaucus.*—Coasts and Islands of Arctic Europe (Arctic Asia and America).

Iceland Gull, *Larus leucopterus.*—Greenland and Jan Mayen.

*Kittiwake, *Rissa tridactyla.*—Greenland, Arctic and Northern Europe.

Great Northern Diver, *Colymbus glacialis.*—Greenland, Iceland, Eastern North America.

*Black-throated Diver, *Colymbus arcticus.*—Northern Europe, Novaya Zemlya, Kolguev, Waigatsch. (Siberia.)

*Red-throated Diver, *Colymbus septentrionalis.*—Greenland, Iceland, Northern Europe, Spitzbergen, Kolguev, Novaya Zemlya. (Siberia.)

Red-necked Grebe, *Podicipes griseigena.*—Southern Scandinavia, Russia.

Slavonian Grebe, *Podicipes auritus.*—Iceland, Scandinavia, Russia. (Siberia.)

BLACK-NECKED GREBE, *Podicipes nigricollis.*—North Central to Southern Europe.

*RAZORBILL, *Alca torda.*—Coasts of Iceland and Scandinavia.

*GUILLEMOT, *Uria troile.*—Coasts of Iceland, Bear Island, and Norway.

*PUFFIN, *Fratercula arctica.*—Coasts of Iceland and Norway.

LITTLE AUK, *Alle alle.*—Greenland, Grimsey (Iceland), Spitzbergen, Franz Josef Land, Novaya Zemlya.

In addition, the following are also to be reckoned as Winter Visitors in small numbers from the North :—

*WOOD-LARK, *Alauda arborea.*

CONTINENTAL GOLDFINCH, *Carduelis carduelis.*

SNOWY OWL, *Nyctea nyctea.*

GREENLAND FALCON, *Falco candicans.*

*HEN-HARRIER, *Circus cyaneus.*

*WATER-HEN, *Gallinula chloropus.*

RUFF, *Machetes pugnax.*

*GREENSHANK, *Totanus nebularius.*

*BLACK-HEADED GULL, *Larus ridibundus.*

*GREAT SKUA, *Megalestris catarrhactes.*

POMATORHINE SKUA, *Stercorarius pomatorhinus.*

*ARCTIC SKUA, *Stercorarius crepidatus.*

*LITTLE GREBE, *Podicipes fluviatilis.*

APPENDIX IV.—LIST OF BIRDS OF PASSAGE

The following species occur annually in spring and autumn as birds of passage in the British Isles. The SEASONAL DISTRIBUTION given indicates whence they may have come, and whither they may be going when traversing our shores :—

GREY CROW, *Corvus cornix.*

Summer.—Northern Continental and Central Europe.

Winter.—South-west Europe, North-west Africa.

ROOK, *Corvus frugilegus.*

Summer.—Northern and Central Europe.

Winter.—Western and Southern Europe.

STARLING, *Sturnus vulgaris.*
Summer.—Northern Continental and Central Europe.
Winter.—Southern Europe.

CHAFFINCH, *Fringilla cœlebs.*
Summer.—Northern Continental and Central Europe.
Winter.—South-western Europe, North-western Africa.

BRAMBLING, *Fringilla montifringilla.*
Summer.—Northern Continental Europe.
Winter.—Western and Southern Europe.

SISKIN, *Spinus spinus.*
Summer.—Northern Continental Europe.
Winter.—Southern Europe, Northern Africa.

TWITE, *Acanthis flavirostris.*
Summer.—Northern Continental Europe.
Winter.—France and South Central Europe.

MEALY REDPOLL, *Acanthis linaria.*
Summer.—Iceland, Northern Continental Europe.
Winter.—Western and South Central Europe.

TREE-SPARROW, *Passer montanus.*
Summer.—Northern and Central Europe.
Winter.—Western Europe.

REED-BUNTING, *Emberiza schœniclus.*
Summer.—Northern Continental Europe.
Winter.—Southern and Western Europe, Northern Africa.

YELLOW BUNTING, *Emberiza citrinella.*
Summer.—Northern Continental Europe.
Winter.—Western and Southern Europe.

ORTOLAN BUNTING, *Emberiza hortulana.*
Summer.—Scandinavia to Arctic circle.
Winter.—? Northern Tropical Africa.

SNOW-BUNTING, *Plectrophenax nivalis.*
Summer.—Northern and Arctic Europe, Iceland.
Winter.—Western and Southern Europe, Northern Africa.

LAPLAND BUNTING, *Calcarius lapponicus.*
Summer.—Northern and Arctic Europe.
Winter.—Western and South Central Europe.

SHORE-LARK, *Otocorys alpestris.*
Summer.—Northern and Arctic Europe.
Winter.—Western and South Central Europe.

SKYLARK, *Alauda arvensis.*
Summer.—Northern Continental and Central Europe.
Winter.—Western and Southern Europe, Northern Africa.

PIED WAGTAIL, *Motacilla lugubris.*
Summer.—South-western Norway.
Winter.—South-western Europe, North-western Africa.

WHITE WAGTAIL, *Motacilla alba.*
Summer.—Iceland, Northern Continental Europe.
Winter.—Southern Europe to North Tropical Africa.

GREY-HEADED WAGTAIL, *Motacilla thunbergi* (*M. flava thunbergi*).
Summer.—Northern Continental Europe.
Winter.—Northern and Tropical Africa.

TREE-PIPIT, *Anthus trivialis.*
Summer.—Northern Continental Europe.
Winter.—Southern Europe to North Tropical Africa.

MEADOW-PIPIT, *Anthus pratensis.*
Summer.—Northern Continental and Central Europe, Iceland.
Winter.—Western and Southern Europe, Northern Africa.

GOLDCREST (Continental race), *Regulus regulus.*
Summer.—Northern Continental Europe.
Winter.—Western and Southern Europe, Northern Africa.

GREAT GREY SHRIKE, *Lanius excubitor.*

Summer.—Northern Continental Europe.

Winter.—Western Europe.

RED-BACKED SHRIKE, *Lanius collurio.*

Summer.—Southern Scandinavia.

Winter.—Tropical, Equatorial, and Southern Africa.

WHITETHROAT, *Sylvia sylvia.*

Summer.—Southern Scandinavia.

Winter.—Tropical and Equatorial Africa.

LESSER WHITETHROAT, *Sylvia curruca.*

Summer.—Southern and Central Scandinavia.

Winter.—Northern and North Tropical Africa.

BLACKCAP, *Sylvia atricapilla.*

Summer.—Southern and Central Scandinavia.

Winter.—Southern Europe to Tropical Africa.

GARDEN-WARBLER, *Sylvia borin.*

Summer.—Scandinavia.

Winter.—Tropical and South Africa.

WILLOW-WARBLER, *Phylloscopus trochilus and P. t. eversmanni.*

Summer.—Northern Continental Europe.

Winter.—North-western to South Africa.

NORTHERN CHIFFCHAFF, *Phylloscopus abietana (P. collybita abietina).*

Summer.—Scandinavia.

Winter.—Southern Europe, Northern Africa

YELLOW-BROWED WARBLER, *Phylloscopus superciliosus.*

Summer—Siberia.

ICTERINE WARBLER, *Hypolais hypolais.*

Summer.—Southern and Central Scandinavia.

Winter.—Tropical and Southern Africa.

SEDGE-WARBLER, *Acrocephalus schœnobœnus.*
Summer.—Scandinavia, Northern Russia.
Winter.—Tropical and Southern Africa.

MISTLE-THRUSH, *Turdus viscivorus.*
Summer.—Southern and Central Scandinavia.
Winter.—Southern Europe, Northern Africa.

SONG-THRUSH (Continental race), *Turdus musicus.*
Summer.—Southern and Central Norway.
Winter.—Southern Europe, Northern Africa.

REDWING, *Turdus iliacus.*
Summer.—Iceland, Northern Continental Europe.
Winter.—Western and Southern Europe, Northern Africa.

FIELDFARE.—*Turdus pilaris.*
Summer.—Northern Continental Europe.
Winter.—Western and Southern Europe, Northern Africa.

BLACKBIRD, *Turdus merula.*
Summer.—Southern and Central Scandinavia.
Winter.—Western and Southern Europe.

RING-OUZEL, *Turdus torquatus.*
Summer.—Scandinavia.
Winter.—Southern Europe, Northern Africa.

REDBREAST (Continental race), *Erithacus rubecula.*
Summer.—Southern and Central Scandinavia.
Winter.—Western and Southern Europe, Northern Africa.

RED-SPOTTED BLUETHROAT, *Cyanecula suecica.*
Summer.—Northern Continental Europe.
Winter.—North Tropical and Equatorial Africa.

REDSTART, *Ruticilla phœnicurus.*
Summer.—Southern and Central Scandinavia.
Winter.—Northern Tropical Africa.

WHEATEAR, *Saxicola œnanthe.*

Summer.—Northern Continental Europe.
Winter.—Tropical and Equatorial Africa.

GREATER WHEATEAR, *Saxicola leucorrhoa* (*S. œnanthe leucorrhoa*).

Summer.—Labrador, Greenland, Iceland.
Winter.—Western Africa.

WHINCHAT, *Pratincola rubetra.*

Summer.—Southern and Central Scandinavia.
Winter.—North Tropical Africa.

SPOTTED FLYCATCHER, *Muscicapa grisola.*

Summer.—Scandinavia, North Russia.
Winter.—South Tropical and Southern Africa.

PIED FLYCATCHER, *Muscicapa atricapilla.*

Summer.—Scandinavia.
Winter.—Northern Tropical and Equatorial Africa.

SWALLOW, *Hirundo rustica.*

Summer.—Northern Continental Europe.
Winter.—Tropical, Equatorial, and South Africa.

MARTIN, *Chelidon urbica.*

Summer.—Northern Continental Europe.
Winter.—Equatorial and South Africa.

SAND-MARTIN, *Cotile riparia.*

Summer.—Northern Continental Europe.
Winter.—Tropical and Equatorial Africa.

SWIFT, *Cypselus apus.*

Summer.—Northern Continental Europe.
Winter.—Tropical, Equatorial, and South Africa.

NIGHTJAR, *Caprimulgus europæus.*

Summer.—Southern Scandinavia.
Winter.—Equatorial and South Africa.

WRYNECK, *Iÿnx torquilla.*
Summer.—Southern and Central Scandinavia.
Winter.—Northern and North-tropical Africa.

HOOPOE, *Upupa epops.*
Summer.—North Central and Central Europe.
Winter.—Northern Tropical Africa.

CUCKOO, *Cuculus canorus.*
Summer.—Northern Continental Europe.
Winter.—Equatorial and South Africa.

SHORT-EARED OWL, *Asio accipitrinus.*
Summer.—Northern Continental Europe.
Winter.—Western and Southern Europe to Southern Africa.

HONEY-BUZZARD, *Pernis apivorus.*
Summer.—Southern Norway, Sweden, North Russia.
Winter.—Tropical and Southern Africa.

MERLIN, *Falco æsalon.*
Summer.—Iceland, Northern Continental Europe.
Winter.—Western and Southern Europe, North Africa.

KESTREL, *Falco tinnunculus.*
Summer.—Northern Continental Europe.
Winter.—Western and Southern Europe.

OSPREY, *Pandion haliaëtus.*
Summer.—Northern Continental Europe.
Winter.—Southern Europe, Africa.

HERON, *Ardea cinerea.*
Summer.—Scandinavia.
Winter.—Western and Southern Europe, Africa to Cape Colony.

GREY LAG-GOOSE, *Anser anser.*
Summer.—Iceland, Scandinavia, Russia.
Winter.—Western and Southern Europe.

WHITE-FRONTED GOOSE, *Anser albifrons.*
Summer.—Iceland, Novaya Zemlya.
Winter.—Western and Southern Europe.

BEAN-GOOSE, *Anser segetum.*
Summer.—Northern Scandinavia and Russia, Kolguev, Novaya Zemlya.
Winter.—Western and Southern Europe.

PINK-FOOTED GOOSE, *Anser brachyrhynchus.*
Summer.—Iceland, Spitzbergen.
Winter.—Western Europe.

BERNACLE GOOSE, *Branta leucopsis.*
Summer.—Greenland, Spitzbergen.
Winter.—Western Europe.

BRENT GOOSE, *Branta bernicla.*
Summer.—Greenland, Spitzbergen, Franz Josef Land, Kolguev.
Winter.—Western and Southern Europe.

WHOOPER, *Cygnus musicus.*
Summer.—Iceland, Northern Continental Europe.
Winter.—Western and Southern Europe.

BEWICK'S SWAN, *Cygnus bewicki.*
Summer.—North-east Russia, Kolguev, Novaya Zemlya.
Winter.—Western Europe.

MALLARD, *Anas boscas.*
Summer.—Greenland, Iceland, Northern Continental Europe.
Winter.—Western and Southern Europe.

GADWALL, *Anas strepera.*
Summer.—Iceland, Russia, Central Europe.
Winter.—Western and Southern Europe, Northern Africa.

SHOVELER, *Spatula clypeata.*
Summer.—Northern Continental Europe.
Winter.—Western and Southern Europe, Northern Africa.

PINTAIL, *Dafila acuta.*

Summer.—Iceland, Faroes, Northern Continental Europe.
Winter.—Western and Southern Europe, Northern Africa.

TEAL, *Nettion crecca.*

Summer.—Iceland, Northern Continental Europe.
Winter.—Western and Southern Europe, Northern Africa.

WIGEON, *Mareca penelope.*

Summer.—Iceland, Kolguev, Waigatsch, Northern Continental Europe.
Winter.—Western and Southern Europe, Northern Africa.

POCHARD, *Fuligula ferina.*

Summer.—Russia, North Central Europe.
Winter.—Western and Southern Europe, Northern Africa.

TUFTED DUCK, *Fuligula fuligula.*

Summer.—Northern Continental Europe.
Winter.—Western and Southern Europe, Northern Africa.

SCAUP-DUCK, *Fuligula marila.*

Summer.—Iceland, Faroes, Northern Continental Europe.
Winter.—Western and Southern Europe.

GOLDENEYE, *Clangula glaucion.*

Summer.—Iceland, Northern Continental Europe.
Winter.—Western and Southern Europe.

COMMON SCOTER, *Œdemia nigra.*

Summer.—Iceland, Northern Continental Europe.
Winter.—Western and South-western Europe, Northern Africa.

VELVET-SCOTER, *Œdemia fusca.*

Summer.—Northern Continental Europe.
Winter.—Western and South-eastern Europe.

GOOSANDER, *Mergus merganser.*

Summer.—Iceland, Novaya Zemlya, Northern Continental Europe.
Winter.—Western and Southern Europe.

RED-BREASTED MERGANSER, *Mergus serrator.*

Summer.—Greenland, Iceland, Northern Continental Europe.
Winter.—Western and Southern Europe, Northern Africa.

RING-DOVE, *Columba palumbus.*

Summer.—Northern Continental Europe.
Winter.—Western and Southern Europe, Northern Africa.

CORN-CRAKE, *Crex crex.*

Summer.—Scandinavia.
Winter.—Tropical and South Africa.

SPOTTED CRAKE, *Porzana porzana.*

Summer.—Southern Scandinavia.
Winter.— Western and Southern Europe, Northern Tropical Africa.

WATER-RAIL, *Rallus aquaticus.*

Summer.—Iceland, Scandinavia.
Winter.—Western and Southern Europe, Northern Africa.

WATER-HEN, *Gallinula chloropus.*

Summer.—Southern Scandinavia.
Winter.—Western and Southern Europe, Northern Africa.

COOT, *Fulica atra.*

Summer.—Norway.
Winter.—Western and Southern Europe.

DOTTEREL, *Eudromias morinellus.*

Summer.—Scandinavia, Novaya Zemlya, Waigatsch.
Winter.—Southern Europe, Northern Africa.

RINGED PLOVER, *Ægialitis hiaticola.*

Summer.—Greenland, Iceland, Spitzbergen, Kolguev, Novaya Zemlya,
Northern Continental Europe.
Winter.—Western and Southern Europe, Africa to Cape Colony.

GOLDEN PLOVER, *Charadrius pluvialis.*

Summer.—Iceland, Faroes, Northern Continental Europe.
Winter.—Western and Southern Europe, Africa to Cape Colony.

GREY PLOVER, *Squatarola helvetica.*
Summer.—North-east Russia, Kolguev, Novaya Zemlya.
Winter.—Western and Southern Europe, Africa to Cape Colony.

LAPWING, *Vanellus vanellus.*
Summer.—Faroes, Southern and Central Scandinavia, Central Europe.
Winter.—Western and Central Europe, Northern Africa.

TURNSTONE, *Strepsilas interpres.*
Summer.—Greenland, Iceland, Kolguev, Novaya Zemlya, Coasts of Scandinavia.
Winter.—Western and Southern Europe, Africa to Cape Colony.

OYSTER-CATCHER, *Hæmatopus ostralegus.*
Summer.—Iceland, Coasts of Northern Continental Europe.
Winter.—Western and Southern Europe, Africa to Senegambia and Mozambique.

GREY PHALAROPE, *Phalaropus fulicarius.*
Summer.—Greenland, Iceland, Spitzbergen, Novaya Zemlya.
Winter.—Western and Southern Europe, Northern Africa.

RED-NECKED PHALAROPE, *Phalaropus hyperboreus.*
Summer.—Greenland, Iceland, Novaya Zemlya, Scandinavia.
Winter.—Seas of South-western Europe.

WOODCOCK, *Scolopax rusticula.*
Summer.—Scandinavia, Russia.
Winter.—Western and Southern Europe, Northern Africa.

GREAT SNIPE, *Gallinago major.*
Summer.—Scandinavia, North Russia.
Winter.—Africa to Cape Colony.

COMMON SNIPE, *Gallinago gallinago.*
Summer.—Iceland, Northern Continental Europe, Central Europe.
Winter.—Western and Southern Europe to North Tropical Africa.

JACK SNIPE.—*Gallinago gallinula.*
Summer.—Northern Scandinavia, North Russia.
Winter.—Western and Southern Europe, Northern Africa.

I.   E

DUNLIN, *Tringa alpina.*

Summer.—Greenland, Iceland, Kolguev, Waigatsch, Northern Continental Europe.
Winter.—Western and Southern Europe to Southern Africa.

LITTLE STINT, *Tringa minuta.*

Summer.—Tundras of Northern Continental Europe, Kolguev, Novaya Zemlya, Waigatsch.
Winter.—Southern Europe, Africa to Cape Colony.

CURLEW-SANDPIPER, *Tringa subarquata.*

Summer.—Tundras of Siberia.
Winter.—Mediterranean, Africa to Cape Colony (Patagonia, Tasmania).

PURPLE SANDPIPER, *Tringa maritima.*

Summer.—Greenland, Iceland, Spitzbergen, Franz Josef Land, Novaya Zemlya, Norwegian Coasts.
Winter.—Western and Southern Europe, North-west Africa.

KNOT, *Tringa canutus.*

Summer.—North Greenland, Tundras of Northern Siberia, Arctic America.
Winter.—Western and Southern Europe to Equatorial Africa, etc.

SANDERLING, *Calidris arenaria.*

Summer.—Greenland, Spitzbergen, Tundras of Siberia.
Winter.—Western and Southern Europe to Coast of Cape Colony.

RUFF, *Machetes pugnax.*

Summer.—Southern Continental Europe, Waigatsch, Central Europe.
Winter.—From Northern Africa to Cape Colony.

COMMON SANDPIPER, *Totanus hypoleucus.*

Summer.—North Europe to Arctic Circle.
Winter.—Southern Europe to Southern Africa.

WOOD-SANDPIPER, *Totanus glareola.*

Summer.—Northern Continental Europe.
Winter.—Tropical and Southern Africa.

GREEN SANDPIPER, *Totanus ochropus.*

Summer.—Northern Continental Europe.
Winter.—From Northern to Tropical Africa.

REDSHANK, *Totanus calidris.*

Summer.—Iceland, Faroes, Northern Continental Europe.
Winter.—Western and Southern Europe to Southern Tropical Africa.

SPOTTED REDSHANK, *Totanus fuscus.*

Summer.—Northern Scandinavia and Russia. (Siberia.)
Winter.—Southern Europe to South Africa.

GREENSHANK, *Totanus nebularius.*

Summer.—Northern Continental Europe.
Winter.—Western and Southern Europe, Northern Africa to Cape Colony.

BAR-TAILED GODWIT, *Limosa lapponica.*

Summer.—Arctic Continental Europe, Western Siberia.
Winter.—Western and Southern Europe to Northern Tropical Africa.

BLACK-TAILED GODWIT, *Limosa limosa.*

Summer.—South-east Iceland, Southern and Central Scandinavia, North Russia, Central Europe.
Winter.—Western and Southern Europe to Southern Africa.

CURLEW, *Numenius arquata.*

Summer.—Northern Continental and Central Europe.
Winter.—Western and Southern Europe to South Africa.

WHIMBREL, *Numenius phæopus.*

Summer.—Iceland, Faroes, Northern Continental Europe.
Winter.—Southern Europe to South Africa.

ARCTIC TERN, *Sterna macrura.*

Summer.—Coasts and Islands of Northern' and Arctic Europe, Greenland.
Winter.—Tropical Atlantic to Antarctic Ocean.

COMMON GULL, *Larus canus*.

Summer.—Northern Continental Europe.
Winter.—Western and Southern Europe, North Africa.

HERRING-GULL, *Larus argentatus*.

Summer.—Coasts of Norway.
Winter.—Western and Southern Europe.

LESSER BLACK-BACKED GULL, *Larus fuscus*.

Summer.—Faroes, Coast of Norway.
Winter.—Western and Southern Europe to West Coast of Tropical Africa.

GREAT BLACK-BACKED GULL, *Larus marinus*.

Summer.—Iceland, Northern Continental Europe, Western Siberia.
Winter.—Western and Southern Europe, Northern Africa, Canaries.

GLAUCOUS GULL, *Larus glaucus*.

Summer.—Coasts and Islands of Arctic Europe.
Winter.—Western and Southern Europe.

KITTIWAKE, *Rissa tridactyla*.

Summer.—Coasts and Islands of Northern and Arctic Europe.
Winter.—Western and Southern Europe, Canaries.

GREAT SKUA, *Megalestris catarrhactes*.

Summer.—Iceland, Faroes.
Winter.—Western and Southern Europe.

POMATORHINE SKUA, *Stercorarius pomatorhinus*.

Summer.—Novaya Zemlya, Tundras of Siberia.
Winter.—South-western Europe to South-west coast of Africa.

ARCTIC SKUA, *Stercorarius crepidatus*.

Summer.—Coasts and Islands of Arctic and Northern Europe, Greenland.
Winter.—Western and South-western European and African Seas.

BUFFON'S or LONG-TAILED SKUA, *Stercorarius parasiticus*.

Summer.—Islands of Arctic Europe, Scandinavia, Greenland.
Winter.—South-western European Seas.

GREAT NORTHERN DIVER, *Colymbus glacialis.*

Summer.—Greenland, Iceland (Eastern North America).

Winter.—West Coasts of Europe, Mediterranean.

BLACK-THROATED DIVER, *Colymbus arcticus.*

Summer.—Northern Continental Europe, Kolguev, Novaya Zemlya, Waigatsch.

Winter.—Western and Central Europe, Mediterranean.

RED-THROATED DIVER, *Colymbus septentrionalis.*

Summer.—Islands of Arctic Europe, Northern Continental Europe.

Winter.—Western Europe, Mediterranean.

RED-NECKED GREBE, *Podicipes griseigena.*

Summer.—Southern Scandinavia, Russia.

Winter.—Western and Southern Europe, North-west Africa.

SLAVONIAN GREBE, *Podicipes auritus.*

Summer.—Iceland, Northern Continental Europe.

Winter.—Western, Central, and Southern Europe.

CHAPTER IV

THE GEOGRAPHICAL ASPECTS OF BRITISH
BIRD-MIGRATION

THE geographical distribution of migratory birds in the British Islands, at any season, obviously depends upon the nature of the particular movement or movements then occurring, or, in other words, upon whence the migrants come and whither they are bound.

In connection with the geographical aspect of migration, it is impossible to over-estimate the value of observations made at islands and rock stations, and other places removed from the usual haunts resorted to by the various species. At such stations to see certain birds is to know at once that they are on migration, for under no other conditions would these particular species be observed there.

The most unsatisfactory of all observations are those made inland. Here individuals of many species moving to other quarters in our islands are most difficult, if not impossible, to distinguish from the native representatives of the same species. In addition, the area and, in many cases, the cover is so extensive that few, very few, of the birds passing through any district come under notice. One never knows what is in the next field or the next bit of cover; while woods are hopeless, it being impossible to ascertain the smaller migrants which are resting in them. Woods in the vicinity of the coast must harbour countless migrants which never come under notice.

THE SOUTH COAST OF ENGLAND.—The south coast of England is the scene, for a prolonged period in both spring and autumn, of the arrival and departure of the great majority of the migratory birds which visit our islands, namely, the Summer Visitors and Birds of Passage.

In spring it receives the vast array of migrants on their arrival from their winter retreats in Southern Europe, or in Northern, Tropical, and Southern Africa. Many of these are the summer visitors which spread themselves far and wide over all parts of our islands ; while many more—the majority—are travellers which journey along our shores to contribute largely to the summer bird-life of vast areas of Continental Europe, Iceland, Greenland, and the Arctic shores, tundras, and islands of the Old World, as far east as Western Siberia.

In autumn the south coast becomes the rendezvous for the retiring summer birds, their ranks now largely augmented by their numerous offspring, which gather there from wide areas, British, Continental, and Arctic, ere they quit our shores to seek their southern and tropical winter homes.

Thus the wide-extending southern coast-line of England is important beyond all others in the British Islands for the visits of migratory birds.

Channel Routes.—During these great arrival and departure movements, the English Channel is crossed by many routes, but there are certain favoured ones, for much migration is observed at both seasons at the Scilly Isles, the Land's End, the Lizard, the Eddystone,[1] Start Point, St Catherine's Point (Isle of Wight [2]), the Nab light-vessel (east of the Isle of Wight), and Beachy Head; and great numbers avail themselves of the shortest passage across the Straits of Dover. These routes have their sources,[3] in spring, at the great stream which rushes along the coast-line of Western Europe, and along the northern coast of France.

The migrants which reach the extreme western section of the south coast and the Scilly Isles and south-west coast of Ireland, do so by quitting the Continental stream flowing northwards along the west coast of France, in the neighbourhood of Ushant. Those which arrive on other sections of our southern coast are derived from the stream which flows eastwards and traverses the French shores of the Channel, giving off branches at various points which pass northwards to the English coast. Other migrants may find their way to England from the opposite shores of the Channel, after an overland flight from various parts of France, in which country their winter has been passed.

Ushant, from its situation, is one of the most important stations for observing bird-migration in Western Europe, and one of these "Studies" is devoted to my experiences during a singularly unfortunate visit paid to the island in the autumn of 1898.

[1] For particulars of the species observed at this important station, see Chapter XVII.

[2] Captain Hadfield (*Zoologist*, 1884, p. 30) says that the main line of flight is witnessed at Freshwater (east of "The Needles"), where more rare species have been obtained than in any other part of the island.

[3] In connection with this and other routes to be mentioned, consult map which forms Plate II.

ROUTES TRAVERSED BY MIGRATORY BIRDS
After Prof. Palmen, Dr Menzbier and W. Eagle Clarke

Plate II

The material originally positioned here is too large for reproduction in this reissue. A PDF can be downloaded from the web address given on page iv of this book, by clicking on 'Resources Available'.

Another favourably situated place for witnessing bird-movements on the opposite shores of the Channel are the Casquets, off Alderney, which lie in the direct course of the stream of migration which traverses the northern coast-line of France, and are visited by great numbers of migrants. In the autumn the birds retrace their flight, and cross the Channel by the same lines as in the spring. On the eastern section of the south coast at that season there is a decided movement of emigrants eastwards towards Newhaven and Dover, whence they cross the Channel for the shores of the Continent, and probably join the stream moving westwards at that season along the north coast of France.

The south coast is traversed in the autumn in a westerly direction by considerable numbers of migrants from the northern and eastern districts of Britain, and also by many migrants from the Continent en route for winter quarters in the southern counties and Ireland. These great western movements along our southern coast-line and its immediate vicinity are renewed in winter, when that season is characterised by periods of unusual severity, the birds fleeing *en masse* before these storms likewise including both British and Continental emigrants, but chiefly the former. There are return movements in spring in an easterly direction, but at that season they are gradually performed, and hence are much less in evidence.

The great majority of the migrants appearing in spring (and departing in the autumn) consists of members of two groups—the Summer Visitors and the Birds of Passage. In addition to these, there are similar movements of the Partial Migrants, and of the birds

evicted by severe climatic conditions (Winter Emigrants). These four groups of migrants have already
been defined (see Chapter III.). The members of the
two last-named groups are usually the first to reappear
in the spring.

MOVEMENTS OF BRITISH SUMMER VISITORS.[1] — On
arriving on the south coast, those birds which intend to
pass the summer in the British Islands at once proceed to
their accustomed nesting haunts. To reach these widely
distributed areas, many pass inland at various points on
the south coast, while others move along the east and
west coast-lines, the courses of rivers being largely
used as highways to the interior. Ireland receives its
summer birds either direct from the south, or from
the south-west of England, after passage across St
George's Channel. The great majority enter Ireland in
spring, and depart from it in autumn, at or about the
south-eastern angle, where stands the Tuskar Rock,
one of the most famous of the Irish stations as an
ornithological observatory.

In seeking and returning from the inland seasonal
haunts, the migrants follow a multiplicity of routes.
The majority of these are mere by-paths which can
only be known in any district to naturalists who
have long resided in it, and who have paid close
attention to the comings and goings of these feathered
visitors.

SPRING ROUTES TO THE NORTH OF BIRDS OF PASSAGE.
—The Birds of Passage are the most numerous of all
the migrants which visit the south coast. We have
learned (p. 35) that they are en route in spring to

[1] Here are included the Partial Migrants which are also Summer
Visitors (see pages 49-50).

wide-reaching areas in the north and east. These far-off
northern destinations are reached by following two great
routes, one of which, the main one, traverses the east
coast of Britain; the other, with several branches,
traverses the west coast and the Irish shores. After
arriving on our southern coast-line, the birds skirt the
Channel shores until they reach its eastern or western
limit, whence they strike a northerly course.

East Coast of Britain.—On the shores of Kent the
south coast migrants about to proceed northwards
receive a great many recruits from the opposite side of
the narrow waters of the Straits of Dover.[1] The east
coast is then followed and the waters of the North Sea
are crossed in a north-easterly direction at many points
between the Humber and the Island of Unst, the
northernmost of the Shetland group, to reach the
Norwegian coast. A number of birds, belonging to
comparatively few species, on reaching the Orkneys and
Shetlands, proceed in a north-westerly direction to the
Faroes, Iceland, and Greenland (see p. 82).

Though vast numbers of the migrants journeying to
northern Continental Europe regularly visit Orkney and
Shetland, yet there is a considerable falling off in the
numbers of a few species, especially among the Hirundinidæ (Swallows, Martins, and Sand-Martins), the main
body of which evidently fly across the North Sea ere
these northern archipelagos are reached. On the other
hand, the ranks of the travellers along our eastern
shores are reinforced at these northern archipelagos by a

[1] At the Varne lightship, situated in the mid waters of the Straits of
Dover, migrants are observed proceeding from the French coast to that of
Kent by a flight from S.E. to N.W. in spring, and in a reverse direction in
autumn.

contingent which has reached the islands by way of the west coast routes, to be presently treated of.

There are a number of famous stations for witnessing these interesting passage movements along the east coast of Britain. Among others, Breydon, Wells, and other places on the Norfolk coast; the shores of the Wash and Humber; Spurn Head, the southern limit of the Yorkshire coast; the Farne Islands; the Isle of May, and the Firths of Forth and Tay; the Montrose Basin, and the numerous isles of the Orkney and Shetland groups.

The points at which the birds depart in the spring and arrive in the autumn, on our eastern shores, are well worth attention, and are best ascertained by a study of the movements in autumn, when the migrants are most in evidence. It has already been stated that these points lie north of the Humber, and between that estuary and the Island of Unst, the northernmost limit of the British area. We are able to determine, with some degree of precision, the southern limit—the one which alone presents difficulties—of these passages across the North Sea. No section of the British coast is so abundantly equipped with light-stations as that which lies between the Humber and the Straits of Dover. Here, in addition to our average number of lighthouses, there is a fleet of lightships, stationed at varying distances off the coast, and most favourably situated for recording these movements between Northern Europe and the eastern shores of Britain. These lightships furnished the British Association Committee on Bird-Migration with carefully kept records for a number of years. These clearly indicate that the great autumn movements of northern species are not observed at these floating observatories, and that the migrants traversing the southern half of the east coast of England, do so *after* arrival on its northern section. In this connection evidence of a particularly important nature is afforded by the records kept at Outer Dowsing lightship, which is anchored some 38 miles off the mouth of the Humber. Here these great movements are not observed: a fact which indicates that the migrants pass to the northward and westward of this vessel in their journeys across the North Sea. As regards northern limits for their arrival from and departure to Scandinavia, our knowledge has been greatly added to by recent investigations, which clearly demonstrate that the great majority of the species travel in considerable numbers via the Shetland Isles. All the movements do not cover the extensive stretch of coast-line indicated, but not unfrequently this is the case. Nor must it be supposed that all of the individuals of the species observed in Orkney and Shetland visit these islands on their journeys to and from the Continent; though many do so, the majority of them cross and recross the North Sea at points further south on our eastern sea-board.

Fair Isle, lying midway between the Orkney and Shetland groups, has proved to be the most remarkable station for the visits of birds of passage at both seasons, and its record is not surpassed by any other locality in the British Isles, and is only equalled by that of Heligoland. The results of several years' investigations, carried out by myself at this small and remote isle, will form the subject of one of the subsequent studies.

West Coast of Britain and the Irish Coasts.—These sections of our coast-line are also traversed by great streams of birds on their passage northwards; but the travellers are composed of fewer species, and in most cases of fewer individuals, than those proceeding via the east coast route. The migrants travelling between the British Islands and the Faroes, Iceland, and Greenland, are, however, probably more numerously represented on the western shores.

The western passage movements are far more complicated in their geographical aspects, and hence more difficult to trace than those witnessed on the eastern side of Great Britain. The migrants proceeding northwards along this route, or series of routes, are part of the great hosts which appear on the westernmost section of the south coast of England and the Scilly Isles in the spring. After arrival on our shores they continue their journey northwards; some skirting the west coast of England and Wales, others traversing the east coast of Ireland, and a number taking a middle course, and visiting the Isle of Man.

When the northern limit of the Irish Sea is reached, the migratory streams which have flowed along its eastern and western shores meet in the narrow waters between the Mulls of Galloway and Cantyre on the one side, and the coast of Antrim on the other. The Scottish coast having been reached, complications await the would-be investigator wishful to follow the further courses pursued. Much migration is observed in the numerous bays of the Solway Firth and on the Galloway coast; but on departing from the latter, the main stream travels via the Mull of Cantyre, Islay, Dhu Hearteach Rock, Tiree, and other inner islands. On reaching the importantly situated rock of Skerryvore, a number of the migrants which have not proceeded northwards by way of the

Inner Hebrides, fly towards Barra Head and traverse the outer group to their northern limit at the Butt of Lewis.

This deflection of the two streams of migration to the west is probably due to the extreme irregularity, ruggedness, and barren nature of the coast of the western mainland of Scotland, with its numerous firths ending in *cul de sacs*—characteristics which render it unsuitable as a migration route, inhospitable and uninviting for migrants, and almost impossible for the observer.

There is yet another western route, the westernmost of all. This lies along the Atlantic coast of Ireland, which is possibly reached direct from the north-west coast of France, and not by way of the south-west coast of England. The regular travellers by this Irish fly-line do not, so far as is yet ascertained, belong to many species, but they include the Whimbrel, White Wagtail, Snow Bunting, Wheatear, White-fronted Goose, Barnacle Goose, Golden Plover, Grey Phalarope, Great Northern Diver.[1] On quitting the Donegal coast, the migrants proceed northwards by way of the Outer Hebrides, including their outermost island, St Kilda. Having reached the northern limit of the mainland and the Inner and Outer Hebridean Islands, the majority of the

[1] The probability is that the west coast of Ireland is much more resorted to as a migration route than is generally supposed. Certain islands off the coast, Tearaght in particular, have produced some of the most interesting migratory birds that have been obtained in Ireland, such as the Yellow-browed Warbler, Red-breasted Flycatcher, Short-toed Lark, Golden Oriole, Lapland Bunting, Greater Redpoll (*A. rostrata*), and others. In all, no less than fifty-one species of migrants have, from time to time, been observed there. If a trained observer were to spend a few weeks on one of these islands during the autumn or spring, much light would be thrown on this interesting far-western stream of migration.

migrants proceed towards the Orkney and Shetland
groups, whence they depart for the shores of Norway.
Others fly direct to the north for the Faroes, and thence
north-west to Iceland and Greenland, to reach which,
however, many of the birds also travel by the Orkneys
and Shetlands.

These western routes boast of several most favourably situated stations for observing these interesting
movements to and from the northern regions.   These
include Skerryvore, situated out in the Atlantic, considerably to the south of the Hebridean groups, being
11 miles W.S.W. of the island of Tiree, and 33 miles
S.S.E. of Barra Head, the southern extremity of
the outer islands.   Many migrants pass this lonely
rock, and fortunately one of the head light-keepers,
the late Mr James Tomison, was a deeply interested
and most capable observer.   As the result of nearly
four years' bird-watching there, he tells us[1] that
the birds (he is speaking of the autumn movements)
come direct from Barra Head, and that on reaching
Skerryvore they slightly alter their course to proceed
due south—a course which would carry them to the
north coast of Donegal.   From Mr Tomison's experience it would seem, as we should expect, that the birds
travelling by the inner islands pass to the east of this
station, for, with the exception of Skylarks seen on
several occasions coming from Tiree, no migrants from
that direction came under his notice.

The outermost fringe of the western stream passes,
as we have seen, along the Atlantic coast of Ireland,
and in its more northerly course touches St Kilda,
where the number of species occurring is quite remarkable

[1] See *Annals of Scottish Natural History*, 1907 pp. 20-31.

for such an out-lying station. Considering their far west position and their insignificant size, the Flannan Islands, a small uninhabited group almost within sight of St Kilda, are also visited in both spring and autumn by many migratory birds. I have spent autumn bird-watching holidays in both these remote groups, and my experiences will be related in the following studies, where a complete account of their migratory birds will be given.

Sule Skerry, an islet situated in the Atlantic 35 miles west of the Orkney island of Hoy, and Foula, the most westerly of the Shetland Isles, are also visited by many birds during the seasons of passage. An account of the former as a bird-station forms the subject of a subsequent chapter.

Other instances of the far west course followed are afforded by the fact that a few Swallows and Fieldfares annually visit the Faroes, presumably on passage to Scandinavia, since they are not summer visitors to Iceland [1] or Greenland.

MIGRATION BETWEEN THE BRITISH ISLES, FAROE, AND ICELAND. — Though the great majority of the birds of passage travelling by the east and west coasts are en route for Northern Europe, yet a considerable number of those which traverse both these great highways are proceeding to and from Iceland. This island is the summer home of a number of birds which belong to essentially Old World species, such as the White Wagtail, Wheatear, and Whimbrel ; [2]

[1] See the special chapters devoted to the histories of these species as British migratory birds.

[2] Also Meadow-Pipit, Redwing, Grey Lag-Goose, Pink-footed Goose, Teal, Common Scoter, Goosander, Water-Rail, Golden Plover, Snipe, Redshank, etc.

and our islands lie in the course of their migrations, when they are proceeding from, and returning to, their European or African winter quarters.   It is probable that the majority of them make their passages along our western shores, for important movements of Red-wings, Wheatears, White Wagtails, and Whimbrels are observed on the west coast of Great Britain and the Irish coasts (both east and west) when they are not observed elsewhere with us.   That many of them visit the Orkneys and Shetlands, and some of them the east coast of Britain on their journeys is manifest from the abundance of the large form of Wheatear (*S. leucorrhoa*), which in summer is confined to Greenland and Iceland. This bird occurs at Fair Isle, and in a lesser degree on the mainland shores of the North Sea, in both spring and autumn.   No doubt many other migrants to and from the north-west also proceed along our eastern sea-board, but being of the same species as the travellers from the northern portions of the Continent, it is impossible to distinguish them.

In instituting a comparison between the east and west coast routes, we find that the birds of passage travelling to and from Northern Europe are more numerous in kinds, and very much more so in individuals, on the east side than on the west.   Many species that traverse our eastern coast-line in vast numbers, and others that do so less abundantly, but yet regularly, occur on the western coasts, either in much smaller numbers or only as occasional or rare visitors.   The following are among the migrants which are less frequent or actually rare on the west coast :—the Redstart, Tree-Pipit, Hedge Accentor, Red-backed Shrike, Pied Flycatcher, Wryneck, Grey Crow, Pink-footed Goose, Dotterel, Great Snipe, Ruff, Green Sandpiper, and Dusky Redshank; while such species as the Blue-throat, Barred Warbler, Great Grey Shrike, Shore Lark, and Honey - Buzzard, are almost unknown in Ireland. The Goldcrest, Swallow, Martin, Sand-Martin, Ring-Ouzel, and Red-backed Shrike are *raræ aves* as Hebridean migrants. On the other hand, the west coast route is probably more used than the east by such species as the White Wagtail, Greater Wheatear (*Saxicola leucorrhoa*), Black-tailed Godwit, the Red-necked and Grey Phalaropes, and Buffon's Skua; and by the regular migrants proceeding to and from Iceland and Greenland.

The lines of migration of several species which summer in Scandinavia lie to the eastward of our islands, and some of them are doubtless overland routes to and from the south and east. These remarks refer to such birds as the Red-throated Pipit, Icterine Warbler, Broad-billed Sandpiper, Temminck's Stint, Crane, and others.

MIGRATION BETWEEN THE SOUTH-EAST COAST OF ENGLAND AND THE COAST OF WESTERN CENTRAL EUROPE: THE EAST AND WEST ROUTES.—The southern section of the east coast of England is the scene of the arrival in the autumn, and departure in the spring, of a very considerable number of bird-travellers. These form a group of migrants of a particularly interesting description, consisting of certain species which cross the southern waters of the North Sea by a more or less direct east-to-west flight in the autumn, and by move-ments in a reverse direction in the spring. They are Central European birds, and in the autumn are

either on passage to lands beyond our shores, or are
bent on passing the winter with us. Day after day in
late September, and throughout October, this stream of
migration, which is often of great volume, is observed
sweeping past the numerous lightships stationed off the
coast between the Straits of Dover and the Wash,
towards the coasts of Kent, Essex, Suffolk, and
Norfolk. It has its centre on the Essex coast and
at the mouth of the Thames, and towards these the
migrants proceed by a direct westerly course. Off the
Suffolk and Norfolk shores the stream is observed
moving in a north-westerly or north-north-westerly
direction; while off the Kentish coast the course is
south-westerly or south-south-westerly.

These lines of flight across the North Sea seem to
be rigidly adhered to. The voluminous observations
made at the lightships clearly indicate that the birds are
not making for the nearest land, but are steadily
pursuing definite courses. This is an important and
significant fact, and it is fair to draw the conclusion that
the courses have been steadily maintained across the
North Sea. If this be so, and I see no reason to doubt
it, then this vast stream of migration flows from the
mouths of the rivers Maas, Schelde, and Rhine. These
great rivers, too, are probably the highways along
which this feathered stream passes to reach the Dutch
coast in the autumn (and up which it proceeds in the
spring) from wide areas in Western Central Europe.
The species travelling along this route also lend coun-
tenance to this view, for there do not appear to be
any essentially northern birds among them: all are
typical natives of Central Europe—Grey Crows, Rooks,
Jackdaws, Starlings, Chaffinches, Greenfinches, Tree

Sparrows, Skylarks, Meadow Pipits, Mistle-Thrushes, Wheatears, Goldcrests, Swallows, Martins, Lapwings, Ringed Plovers, Blue-headed Wagtails, Black Redstarts, Golden Orioles, Hoopoes, Common Buzzards, Black Terns, Avocets, Spoonbills, and, in winter, the Bittern also pass, though in smaller numbers; and more rarely still the Firecrest, Richard's Pipit, Little and Baillon's Crakes, Little Bustard, and Stilt. The most numerous of the migrants travelling along this route are Rooks, Grey Crows, Starlings, Skylarks, Tree - Sparrows, Chaffinches, and Lapwings.

On reaching our shores many of these immigrants move up the Thames valley to reach particular winter quarters in south central England. Others, on the same errand bent, skirt the coast northwards (which they—the Skylarks and Grey Crows in particular—sometimes do as far as the Tees), proceeding inland at various points as they go. Those forming the left wing, and they are a considerable contingent, pass westward along our channel shores to reach winter quarters in southern and south-western England and in Ireland, while others doubtless cross towards the shores of France, en route for more southern lands.

These movements from the east set in about the middle of September, reach their maximum in October, and continue at intervals until mid November. They are renewed on the part of Skylarks, Starlings, Thrushes, Lapwings, etc., during the winter, but only when exceptionally severe weather prevails in Central Europe, and the birds then chiefly pass westwards along our southern shores in search of retreats in the west, including the Scilly Isles and Ireland. Many wild-fowl seek eastern England by an east to west flight, when their haunts on the opposite side of the North Sea are sealed by ice.

There are some remarkable features associated with these east to west autumn passages :—(1.) They are frequently observed for several, and sometimes many, consecutive days.[1] (2.) They take place chiefly during the daytime. (3.) Not a few, probably most, of these migrants are proceeding to winter quarters in our islands, which are situated in latitudes *north of their summer haunts*—a most singular circumstance, but one which is to be explained by the remarkably mild temperature of our winters, the winter isotherm for England being the same as that for western and southern France. (4.) On some occasions those birds which are proceeding westwards or northwards actually cross the course of the coasting migrants moving southwards along our eastern sea-board at the same time.

The return movements in spring towards the east are witnessed, along the same lines of flight, from the middle of February to the middle of April.

In the autumn of 1903, I spent nearly five weeks in the Kentish Knock light-vessel, which lies 33 miles off the coast of Essex and about the centre of this interesting stream of migration, in order to investigate the various conditions under which these east to west movements were performed, and with a view to ascertaining the species participating in them. The results were highly satisfactory, and these, along with my singular experiences, are related in Chapter XVIII. (Vol. II.). Other information will be found in the studies devoted to migrations performed by the Rook, Starling, and Lapwing.

[1] See the history of the migrations performed by the Skylark.

There are other lines of flight between the south-east coast of England and the Continent, especially at its extreme limit.   This complication of routes and cross routes at the neck of the Channel is, no doubt, to be accounted for by the contiguity of the British and Continental areas at the Straits of Dover.   While on board the Kentish Knock lightship, I observed, in the month of September, and during the daytime, Wheat-ears, Martins, Starlings, Skylarks, Pied Wagtails, and Meadow Pipits passing in a south-eastern direction—*i.e.*, from the north coast of Essex towards the French coast about Dunkirk; while a little further south, off the east coast of Kent, there are cross movements in progress in an opposite direction at the same period, from the S.E. to N.W.—*i.e.*, from the northernmost shores of the French coast to the coast of Kent—the birds observed being Rooks, Starlings, Skylarks, Chaffinches, Tree Sparrows, and Swallows.

At the Varne lightship, which is stationed in the mid-waters of the Straits of Dover, Rooks and Starlings are sometimes recorded in spring as moving from S.S.W. to N.N.E., and Swallows from S.E. to N.W.; and in autumn, Swallows and Meadow Pipits leaving England pass to the S.E., and Starlings, Larks, and other "small birds," to the W. and N.N.W.   Mr West, the Master of the North Goodwin lightship, remarks that when a great number of birds pass from E. to W. or N.E. to S.W. in the autumn, bad weather generally follows.

MOVEMENTS OF BRITISH EMIGRANTS.—The autumnal movements of the departing summer birds are simple in their geographical aspects, so far as the eastern portion of Britain is concerned; but, as in the spring, it is not so as regards the western coasts. The first movements from the summer haunts are towards the coasts, which are reached by many local routes, especially along the courses of rivers and their tributary streams.

When the east coast is reached, the emigrants skirt its shores and their vicinity, proceeding southwards usually by easy stages, and gathering strength as they advance, particularly at estuaries and river-mouths, such as the Tay, Forth, Humber, Wash, Thames, etc. They finally leave our southern shores by the same routes along which they travelled to reach them in the spring.

The western emigratory movements are not so simple in their geographical aspects. Here we have the Hebridean and other isles, Ireland, and the Isle of Man, and the numerous firths, sounds, and estuaries of the mainland exercising varied influences. The routes followed by these departing summer birds are the same as those already indicated for the spring movements of the birds of passage, but the direction is the reverse, —*i.e.*, to the southwards. The Outer Hebridean summer birds move towards the north-west coast of Ireland, some of them by way of Skerryvore, while others even traverse St Kilda. Those leaving the inner isles proceed by Dhu Hearteach and Islay, towards the mainland at Cantyre, whence they probably travel along both sides of the Irish Sea. The eastern section of this migratory stream skirts the west coast of Britain, and receives in its course considerable tributaries from the Clyde basin, Galloway, the Solway, the Isle of Man, Wales, and the Bristol Channel. From the Pembroke coast southwards very important additions are received from Ireland. Finally the south-west coast of England is reached, between the Scilly Isles and Start Point (partly by an overland flight across Devonshire), whence the travellers cross the Channel to reach their various winter retreats.

The main body of the birds which are summer guests in Ireland, and those which have traversed her eastern shores on leaving the Scottish coasts and isles, take their departure from the Wexford coast, and cross St George's Channel, for the south-west of England, as has just been mentioned. A more western stream, on quitting the sister isle, probably flows direct towards the coast of Brittany, though it may lap the shores of Scilly and the Land's End.

While traversing our coasts, the migrations of these departing summer birds are much in evidence, for many of them are performed leisurely and during the day-time, the birds feeding and resting as they go. Many, too, pass along them in the night-time, and it is then that the great majority of the travellers quit our shores for their southern winter-homes. These remarks apply equally to the migrations of the birds of passage which have arrived from the north and east, and are likewise moving along our shores at this season to reach their more southern retreats. These two sets of migrants undoubtedly often join forces in September and October, and journey southwards in company.

Some account of these autumn emigratory flights across the Channel will be found in the chapter wherein I relate my experiences as a bird-watcher on the Eddystone.

In addition to the emigratory movements of our summer visitors for their southern winter retreats, there are autumnal migrations from Britain to Ireland on the part of Starlings, Skylarks, Chaffinches, Greenfinches, and sometimes Thrushes, by an east to west flight across the Irish Sea. Anglesey and Rockabill (off the north coast of co. Dublin) are the main points at which these departures and arrivals have been recorded. There are also migrations at the same season from the Rhinns of Islay and Mull of Galloway to the north-east coast of Ireland, on the part of Starlings, Skylarks, and Thrushes. The object of these movements to the west is, no doubt, to reach accustomed winter quarters in Ireland.

AUTUMN ARRIVAL AND PASSAGE MOVEMENTS.—The winter visitors and autumn birds of passage often come to us from the same areas at identical periods, and are composed, with few exceptions, of the same species; in fact they are, in most cases, travelling companions until the British Isles are reached.

These numerous migrants reach our shores from Northern Europe, Western Siberia, Iceland, Greenland, and Central Europe, by the same routes as those followed in the spring, but in the reverse direction; the southern portion of the east coast, south of the Humber, being reached after the birds have arrived on the northern section; and the west coast and the Irish shores by way of the Shetland, Orkney and Hebridean Islands.

On their arrival on our coasts, those of the migrants which are winter visitors move inland at many points to spread themselves over our islands, many reaching the west coast districts after overland passages across the mainland of Britain, while those which are birds of passage, proceeding to winter retreats beyond our shores, traverse the east and west coasts, following the same routes southwards as they traversed when on their way northwards in spring.

The autumn movements of the winter visitors and birds of passage which arrive on the south-east coast of England from Central Europe have already been fully described.

It is desirable that some allusion should here be made to the supposed intermigration between Britain and Heligoland. Much prominence was given in the annual reports issued by the Committee appointed by the British Association, and in Herr Gätke's book, *Die Vogelwarte Helgoland*, to an intermigration between the famous island off the mouth of the Elbe, and the Yorkshire and Lincolnshire coasts, by direct east to west autumn movements. Herr Gätke most obligingly communicated the details of the bird-movements observéd at Heligoland for four of the years (1883-1886) during which the inquiry was being prosecuted over the British area. The two sets of data were carefully examined and compared by me, with the result that I was not able to find any indications whatever that would warrant such a conclusion; nor have any observations been forthcoming since that date in support of such a contention. It is not impossible nor improbable that birds may occasionally cross the North Sea in the latitude of Heligoland, but our present knowledge compels us to believe that such flights must be regarded as exceptions, and not as the rule. This subject was discussed by me in the *Report of the British Association* for 1896, p. 457.

BRITISH INLAND ROUTES.—Though much remains to be ascertained regarding the inland routes used by birds within the British area, yet we know more than enough to enable us to aver that there are no important overland fly-lines for birds of passage such as traverse the Continental land-masses. That this should be so is not at all surprising. Our area is longitudinally too circumscribed to render them necessary, for our coast-lines are near at hand, and these birds keep to them and their immediate neighbourhood with remarkable pertinacity. As an indication of the coasting pro-clivities of these migrants, it may be remarked that all the species which are essentially birds of passage (such as the Bluethroat, Curlew Sandpiper, etc., etc.), are of rare or exceptional occurrence at inland localities in our islands. The Scottish birds of passage include a number of species which are breeding birds in England and in Scandinavia, but not in Northern Britain (such as the Lesser Whitethroat, Red-backed Shrike, Wry-neck, Ruff, and others), and these are practically con-fined to the Scottish coast-line and its neighbourhood when passing northwards in the spring and southwards in the autumn. The occurrence, inland, of any of the species which visit us on passage only is more or less rare, and certainly exceptional.

There are, however, many highways of minor import-ance which deserve some consideration, though our knowledge of most of them is of a very imperfect nature. These are the overland courses followed by the summer and winter visitors when proceeding to and retir-ing from their seasonal haunts on the mainland of Great Britain and Ireland.

After arrival on our shores in spring and autumn many of these immigrants proceed to their accustomed haunts along natural and convenient highways, and, the summer or winter o'er, again seek the coast-line. In the case of some species, to which reference will presently be made, the routes followed are determined by the nature of the food-requirements of the travellers. To others, however, influences of a topographical nature appear to be the most important. Many of the birds after arrival seek estuaries and the mouths of rivers, and from these trace up the course of the main stream and its tributaries, and thus spread themselves over wide areas; some proceeding to their very sources to reach the higher ground, the moorlands, and the fells.

These highways and byeways can only be known to those local naturalists who have for some years given close attention to the subject, and, as yet, comparatively little has been placed on record regarding them.

The Humber Basin and Yorkshire, etc.—My personal experiences of such routes are confined to Yorkshire, in which county one series of inland highways is known to me, namely, that which leads from the estuary of the Humber, by way of its river systems, to the west and north-west of the county. From this source a vast area receives many of its summer birds in spring, and to the Humber they return in the autumn, their numbers then greatly increased by their offspring, to take their departure southwards.

A notable example is the Yellow Wagtail (*Motacilla rayi*). This conspicuous species is extremely abundant in spring and summer, amid the grasslands and fields bordering the upper valleys of the Wharfe, Swale, Ure, Aire, and their tributaries, and forms quite a feature in their bird-life. So far inland are some of these haunts resorted to by this handsome bird, that they are very much nearer the western sea-board than the east coast, and yet the Pennine chain is not crossed, though some *may* reach Ribblesdale, where the species is common, from the east by way of the " Aire gap,"

which is below 500 feet in elevation.   Towards the end
of August, and during the first half of September, the
return takes place, and the migrants on reaching the
Humber move eastwards along the shores of the estuary
to reach the coast of the North Sea, whence they take
their departure for their winter quarters.   These retiring
movements are sometimes performed simultaneously from
considerable areas, and may then be said to take the form
of a "rush."   Thus, on the 23rd of August 1885, I saw
literally thousands on the north shore of the estuary and
its immediate vicinity, all working their way towards
Spurn Head.   Most of these, perhaps all, took their
departure the same night, for on the following day
comparatively few were seen, and these may have been
fresh arrivals.   These emigratory movements, as a rule,
extend over many days, and are most interesting to
witness.

The southern shore of the Humber is also traversed
by large numbers of these Wagtails, many of them
proceeding to and from the Trent valley, where they are
also very abundant during the nesting season.   Similar
movements also proceeding to and from the Yorkshire
dales by way of the Humber are observed on the part of
such species as the Wheatear, Redstart, Whitethroat,
Ring-Ouzel, Pied Flycatcher, Sandpiper, etc.

In the autumn considerable numbers of Starlings,
Skylarks, and Lapwings, from the south-east enter the
Humber at its mouth, and pass along its southern
shores, proceeding westwards.   These immigrants appear
with great regularity, and have come much under the
notice of Mr Caton Haigh.

The Trent Valley, already alluded to, is a much-
used highway, according to Mr Whitlock (*Zoologist*,

1891, pp. 178-9), who describes the effect of its use as very apparent in Nottinghamshire. He remarks that its most interesting feature is the fact of its being traversed in spring, when amongst the other birds using this route are the Dunlin, Common Sandpiper, Redshank, Sand-Martin, and Yellow Wagtail. In his *Birds of Derbyshire*, this same author (pp. 15 *et seq.*) states that the Trent valley is "extensively used as a fly-line of birds travelling to and from their breeding grounds in the north, and naturally causes this portion of the county (*i.e.*, that which borders the Trent) to be richest in bird-life, both as regards numbers and species."

The Wash.—The late Lord Lilford (*Zoologist*, 1891, p. 52) states his opinion that the valley of the Nene, from the Wash as far up as Thrapston, is certainly a much-used route of migration, but he believed that the majority of the autumn migrants left the valley somewhere above that town, and struck across the country for the eastern affluents of the Severn. Unfortunately, his lordship did not name any of the species using this route.

Mr O. V. Aplin informs me that he is of opinion that certain winter visitors reach Oxfordshire from the Wash via the Northamptonshire valleys. Also, that there is an important fly-line crossing Oxfordshire from north-east to south-west between the Wash and the Bristol Channel. This is much used by Gulls in the autumn, and in a lesser degree by some Waders.

Thames Valley.—This is another much-used route to and from the interior of southern England. On 18th October 1903, I traced great numbers of Starlings and Skylarks (which were then crossing the North

Sea) from the Kentish Knock lightship, 33 miles off the coast of Essex, as far as Southend, at which place I disembarked. The birds, however, were still trooping up the Thames valley in considerable numbers when I left the ship.

Numbers of Rooks, Starlings, Skylarks, Fieldfares, and other species have been noticed in October at Bermondsey, making their way in a north-westerly direction. On some occasions, as on the 22nd October 1896, vast numbers came under notice, phalanx succeeding phalanx in regular military order, without any intermingling. This great flight lasted for half-an-hour, during which, over thirty flocks were counted, each numbering thousands of birds of the above-mentioned species. Regarding the spring, Mr O. V. Aplin tells me that the summer migrants reach Oxfordshire by this route.

Devonshire.—There is an interesting cross-country route from Barnstaple Bay, on the north coast, by way of the Torridge and the Teign valleys to the south coast of the county. This "short cut" is much followed in the autumn by migrants seeking winter quarters across the Channel. For this information I am indebted to Mr A. S. Elliot.

North Wales.—Mr H. E. Forrest, in his *Fauna of North Wales* (p. 65), indicates an overland route up the Wye valley, followed by birds to reach their summer quarters, some of which are in the southern portion of Montgomeryshire. He also mentions that in Shropshire the Severn valley is much traversed by migrants.

Lakeland.—The late Rev. H. A. Macpherson informed me in the year 1900 that he saw no reason to depart from the views he had expressed regarding cross-country migration in the *Birds of Cumberland* (p. 12). In that work three lines were indicated by which migrants arrived in the county in the autumn, and which appeared to be reverted to by many species on their spring emigration. These are: (1) a line starting from Berwick, and, passing from north-east to south-west, so as to culminate on the Solway Basin; (2) a line from Tynemouth, which, following the rivers Tyne and Irthing, would meet the first line on the Solway; (3) a line from the Durham coast, which, passing through Weardale or Teesdale, would enter Cumberland, near Alston, thence trending south-west to the Ravenglass and Duddon estuaries.

Solway.—The late Mr Service favoured me with the following interesting information, the result of particularly ripe experience, relating to overland migration in the Scottish section of the Solway area. There is, he stated, an extremely well-marked line of migration to and from the Solway in the direction of the Clyde—a short cut from the Ayrshire coast, and the most important route to and from the Solway for waders and swimmers, which follow it in great numbers. The route northwards is up the Nith, across the hills near Cumnock, and then straight to the shore north of Ayr. There is also a direct east-to-west route, and *vice versa*, according to the season, taken by Fieldfares, Redwings, Skylarks, Snipe, Woodcocks, etc. To be quite correct, it is a little north of east in the autumn, and a little to the south of west in the spring.

Forth, Clyde, and Solway.—Mr William Evans has supplied me with the following statement relative to the passage of certain species between these areas:—Many birds undoubtedly cross Scotland every autumn by way

I. G

of the broad stretch of (for the most part) low country separating the upper estuaries of the Forth and the Clyde (of this there is ample evidence), and also by a longer southerly route from mid-Forth to the Solway. Leaving Forth either by the wide gap between the Moorfoot and the Pentland Hills, or by one of the western Pentland passes, the travellers taking the southerly passage soon reach the head-waters of Tweed and Clyde, and thence by way, it is inferred, of the valley of the Nith or the Annan Water down to the Solway. Oyster-catchers, Curlews, Whimbrels, Ringed Plovers, Dunlins, and Common Terns are to be seen every year between the end of July and the end of September, but chiefly perhaps in August, proceeding on the Forth to Solway flight. The reservoirs along the north-western base of the Pentlands, at which the birds often halt, are good points for observation. At Crosswood reservoir, for example, Mr Evans has watched all the above-named species depart, their course being almost due south, which would take them through the hills by the sources of the Medwyn to the head of Clyde. At Elvanfoot, which is only a short distance from the Clyde and Solway watershed, he observed Oyster-catchers, etc., passing in a southerly direction· in September 1900. They have also been noted near West Linton, and (Terns included) at Stobo on Tweed. Besides the Waders and Terns, which are no doubt chiefly our native birds, Fieldfares, Redwings, Wild Geese, etc., have also been noticed passing by these routes to their winter quarters west and south of Forth, but at a somewhat later date. The return movement in spring, though doubtless likewise of annual occurrence, has been less observed. It is not known that any of the

true birds of passage traverse these routes, but Mr
Evans considers it highly probable that a number of
summer visitors—Wheatears, Ring-Ouzels, Redstarts,
Pied Flycatchers, for instance—reach their breeding
grounds in the central parts of the southern section of
Scotland by way of them.

Forth and Clyde.—Mr Harvie-Brown has indicated
that there is a much-used overland route between Forth
and Clyde. He often hears migrants passing during
the night over Dunipace, and proceeding from east to
west during the autumn, and towards the east and
north-east in spring.

Ireland.—Mr Allan Ellison (*Zoologist*, 1885, p. 18)
mentions a north-east to south-west overland route
from Co. Wicklow, via Shillelagh, to the south-west of
Ireland, used by Starlings, Fieldfares, Redwings, Skylarks, and Golden Plovers.

The Shannon and its lakes, Mr Ussher (*Irish
Naturalist*, 1905, p. 125) tells us, afford a north-to-
south route, while another very easy route for wild-fowl
passing from Killala Bay to Galway Bay, is by way of
lakes Corrib, Mask, and Conn.

Other Routes.—The overland highways of some
species are determined by the special nature of their food
or by a predilection for certain peculiar haunts—con-
siderations which exert not a little influence on the
courses followed by some migrants when passing to and
from their British nesting haunts—the Greenshank,
Dotterel, Whimbrel, among others.

Some Greenshanks (*Totanus nebularius*), probably
on quitting their nesting grounds in the Scottish High-
lands, move southwards overland, proceeding from
one sheet of water to another, or from river to river.

In the autumn a few annually visit suitable localities in central England, where they may be observed, chiefly in September. When I resided in Yorkshire, I used to see several each year on the margins of a series of moorland reservoirs in the heart of the county, where they lingered for some weeks, apparently being attracted by the suitable feeding grounds afforded. South of the Humber, this bird appears to be somewhat irregular in its inland visits, and probably travels further south by way of the coast-line, which it may reach by way of the Yorkshire rivers which debouch into that great estuary. Other species which follow similar routes are the Dunlin, Common Sandpiper, Snipe, Common Tern, and certain Ducks.

The Dotterel (*Eudromias morinellus*), when en route to and from its British summer quarters amid the mountains of north-west England and of Scotland, traverses the more elevated inland districts, and is observed on the Chiltern Hills, and the fells of the Pennine chain. Here it was formerly observed in flocks, especially in the spring, but unhappily a very marked diminution in their numbers has taken place in recent years. This diminution corresponds with the decrease of this bird in its British nesting haunts, and is an interesting though melancholy fact, since it clearly indicates that the migrants traversing the high ground of central England were (and are) the birds which breed in our islands. The Dotterel is one of the comparatively few British migratory birds which is more in evidence during the spring movements than in those of the autumn — a somewhat remarkable fact, for the bird is naturally more abundant during the latter season. Other species which affect the moorlands or similar elevated districts in the summer, also traverse similar tracts of country when passing to and fro—among them are the Golden Plover, Curlew, etc.

The Whimbrel (*Numenius phæopus*) is another species which passes overland in spring and autumn, though it apparently does not follow very definite inland routes—a circumstance which is no doubt accounted for by the fact that it seldom alights to feed. It is generally observed passing north or south, according to the season, at a considerable elevation, attention being drawn to it by the well-known and peculiar callnote which is continually uttered. In my experience, gained in central Yorkshire, these birds pass over during the daytime, and usually singly or in pairs.

RACIAL FORMS OF MIGRATORY BIRDS.—In connection with the geographical aspect of bird migration, it is of importance to allude to the racial forms to be found among a number of the species which occur as visitors to our islands and their shores. Much attention has, fortunately, been paid to the study of these geographical forms during recent years, especially by Dr Hartert, who has made many of them known to us, and is treating of them in his excellent work, *Die Vögel der paläarktischen Fauna.*

A knowledge of these geographical races is of the utmost value to those interested in bird-migration, for it enables the student in a number of cases to determine whence certain migrants set out on their journeys to reach our shores—a gain the importance of which it is impossible to over-estimate. A number of these racial forms regularly appear in our islands, either as seasonal visitors or as occasional guests. Among these are Continental or Arctic races of the following species :—Mealy Redpoll (several forms), Bullfinch, Crossbill, Creeper, Goldcrest, Willow - Warbler, Chiff-Chaff, Redbreast, Wheatear, Song - Thrush, Dipper, Great Spotted Woodpecker, Barn-Owl, etc.

Such is a broad general outline, based on many years devoted to their investigation, of the geographical distribution of birds during their various migratory movements in the British area. It must be left to local observers to fill in the details—they alone have the opportunities for acquiring the necessary special knowledge.

CHAPTER V

ROUND THE YEAR AMONG THE BRITISH MIGRATORY
BIRDS : SPRING

> The Swallow knows her time,
> And, on the vernal breezes, wings her way
> O'er mountain, plain, and far-extending seas,
> From Afric's torrid sands to Britain's shore. —GRAHAM.

THAT migratory birds observe with remarkable regularity the times of their coming and going, has been known since the days of the prophets ; and this fact so much impressed the untutored redskin of the fur-countries, that, in framing his primitive calendar, he named the recurring moons after the migrants whose appearance was synchronous with their advent.   In our own country the arrival of certain well-known species has long been associated with the advent of the seasons, or noted as an indication of their meteorological peculiarities.

The seasonal movements relate to the spring, autumn, and winter.   Summer proper has no place in the calendar of bird-migration, but it is closely approached by the late spring movements, and is actually trenched upon by migrations which from their nature can only be classified with those of the autumn.

The migrations undertaken during each of the seasons have characteristics and peculiarities of their own, for they are performed under different influences and for different purposes.

SPRING.

At no other season do migratory birds attract so much attention, or arouse the same widespread degree of interest as in the spring. This is not surprising, for time out of mind their appearance among us has been regarded as the harbinger of a glad season, and an omen of the passing of one which has become drear. They are most welcome visitors, too, for in their ranks are to be found some of the most famous songsters, such as the Nightingale, Blackcap, and a host of others, which contribute with much acceptance to the joyousness of the season, and without which the pageant of spring would indeed be a spiritless show, for it has been well said by Cowper—

Nor country sights alone, but country sounds,
Exhilarate the spirit.

All the spring movements in the British Isles and elsewhere north of the equator, varied though they may seem to be, are undertaken for the same object, namely, to return to summer haunts, whether those haunts be close at hand or in lands far removed from where the winter was spent.

At no other period are the migrations performed under such all-engrossing conditions. It is the season which brings in its train the revival of the sexual activities, with their irresistible incentives to repair to the nesting retreats : it is the season of love-making, nest-building, and the rearing of families. Need one be surprised that the migrants proceed on hurried wing to reach these hallowed scenes? This race for breeding haunts is one of the main characteristics of the spring movements, and is especially in evidence on the British shores, for our isles form one of the last stages in the journey of a vast number of birds on passage to their native lands.

> The birds of passage transmigrating come,
> Unnumbered colonies of foreign wing,
> At Nature's summons. —MALLET.

The great majority of these voyagers do not tarry with us: indeed, it is a case of "here to-day, off to-morrow." On this account, their visits afford comparatively few facilities for observation, and hence some species and vast numbers of individuals entirely or largely escape notice—a fact which accounts for the supposed non-appearance of certain birds in the spring which are regularly observed in the autumn.

Another characteristic of the spring is that the males, the more ardent suitors, of most species, travel in advance of the females and arrive at the nesting quarters some days, it is said in some cases even weeks, before their consorts — a circumstance which affords additional evidence of the enthralling nature of the season.

The times of the coming of birds of identical species to our shores, depends upon whether the individuals are bent on spending the summer with us, or are birds en route for distant countries. The first to arrive are, undoubtedly, birds seeking their nesting haunts within the British Islands; the latest are those on their way to summer homes far to the north of us. There are usually many intermediate haunts between southern England and the Arctic countries, and hence birds of the same species arrive on our shores in a series of flights, and at intervals covering in many cases some weeks.   To take a few instances culled from the records in my possession for the spring of 1908 : I find that movements of the Common Wheatear, from the date of its first appearance on the south coast to the passing away of the last of the passage birds from our islands, covered 61 days; those of the Ring-Ouzel, 49 days; of the Willow-Warbler, 65 days; of the Blackcap, 76 days; of the Common Sandpiper, 83 days; and of the Whimbrel, 54 days.   If we knew all, the periods would be longer.   On the other hand, the period covered by the Nightingale, a bird which is a summer visitor to England only, was but 24 days, perhaps less.

As regards the dates of arrival of summer birds, it should be remarked that observations made at inland localities are in a great many instances of little value, since an incomer may remain for a number of days undetected.   Another fact, of the accuracy of which I am convinced, is that not one migrant in several thousands comes under notice immediately on its arrival in the British Isles.

In not a few cases, as we shall see, the birds which traverse our coasts on their way to the northern breeding grounds in May and June do so at a time when their British representatives are either engaged in the duties of incubation or in tending their young.   Song-Thrushes, for instance, pass northwards when the young of our native birds have left the nest, and Golden Plovers when the chicks of their British cousins are well grown.

We may now proceed to consider the various movements performed during the months which constitute spring from the bird-watcher's point of view.

FEBRUARY.—It requires some stretch of the imagina-

PLATE III.

THE ADVANCE OF SPRING
After H. Hildebrandsson

The map illustrates the march of isotherm 48° F., "the climatic beginning of spring," across Europe.

The lines show the fortnightly advance of this spring temperature from the Mediterranean Sea to the Arctic Ocean, and indicate approximately the dates on which the various regions of the Continent became climatologically suitable for the reception of their bird summer visitors.

Plate III

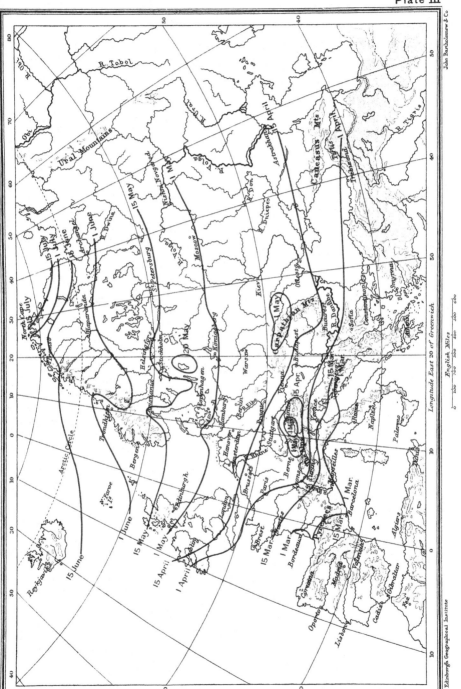

tion to include in the category of spring the wintry month of February. Yet within its allotted days movements take place which from their very nature, namely, the seeking of nesting haunts, are strictly in consonance with those of spring. This month, however, belongs to both spring and winter in its migration aspects; here we are only concerned with those of the former season.

Local Movements.—The earliest of the February movements relate to visits to summer quarters of birds which have passed the winter in British retreats not very far removed from the nesting area. Should the month prove a genial one, from its second week onwards it may witness the return to the upland woodlands of the Song-Thrushes which quitted them during the previous autumn; of the Redshanks, Curlews, Snipe, and Meadow-Pipits to the moorlands; and of the Skylarks and Lapwings to the higher pastures, etc. During the month, too, there are return movements of Pied Wagtails, Meadow-Pipits, Skylarks, Lapwings, Oyster-catchers, etc., to the Orkneys and to northern localities on the mainland; and the Gannets, Kittiwakes, Black Guillemots, Common Guillemots, Razorbills, and other maritime species pay visits to their breeding places.

The experiences of some of these earliest visitors to their summer quarters are generally of an unfortunate nature, especially for the ground-loving species, inasmuch as a return of winter conditions, with their pall of snow or frost-bound lands, is almost certain to follow and to drive them ruthlessly away. Later, in March, and even in April, such a renewal of winter ends in disaster, for then they are loth to quit their haunts, and perhaps their nests, and many perish.

Some Skylarks and Lapwings which have sought Ireland in the autumn, to pass the winter in its milder climate, return across the Irish Sea to their quarters in Great Britain.

Arrival of Earliest Migrants.—Should spring-like periods occur during the latter half of February, the southern coasts of England and Ireland receive, after a passage across the Channel, the first birds to return from their Continental winter retreats.[1]   These belong to that section of our summer birds known as Partial Migrants ; the species seeking again their native land at this period are Song - Thrushes, Mistle - Thrushes, Red-breasts, Starlings, Goldcrests, Skylarks, Pied and Grey Wagtails, Meadow-Pipits, and Lapwings.   These earliest immigrations of the year come chiefly under notice at the light-stations, and I have been furnished with many valuable data regarding them from the advantageously situated watch-tower on the Eddystone.   At ordinary points on the coast they are liable to be overlooked, for the arrivals chiefly take place during the hours of darkness, being timed between 7 P.M. and 6.30 A.M. Starlings, however, have been observed making for the Cornish coast during the forenoon.   In most seasons the return is a gradual one, but occasionally a rush is recorded, as, for instance, at the Eddystone on the night of 19th-20th February, 1903.   On this occasion, from 7 P.M. to 5 A.M., Starlings, Song-Thrushes, Mistle-Thrushes, Skylarks, and Lapwings were passing northwards in numbers, and many came to grief by dashing against the lantern.   This is only one of many similar movements recorded in the returns from this important station.

---

[1] The earliest date known to me for such immigrations is 11th February, and the birds chronicled were Skylarks.

Late in the month numbers of Rooks, Starlings, and Skylarks occasionally arrive from the Continent during the daytime on the south-east coast between Kent and Norfolk, but as these immigrations more properly belong to March, they will be treated of when dealing with the movements observed during that month.

Associated with the return of these British native birds from the south are the similar movements of those birds (Fieldfares, Redwings, Blackbirds, etc.) which had fled our country in the winter owing to the severity of the weather.

Exceptional Arrival of Summer Visitors.[1] — In February, strange to relate, a few species which are essentially summer visitors to our islands, have been known to be so very indiscreet as to appear upon our, as yet, inhospitable shores, it is to be feared in most cases with disastrous results. The visits of these deluded birds must be regarded as phenomenal, and are perhaps to be accounted for by an outburst of exceptionally fine weather in the regions in which they have spent the preceding part of the winter. The following are the species which have been recorded :—Ring-Ouzel, Wheatear, Blackcap, Chiff-Chaff, Tree-Pipit, Swallow, Sand-Martin, Nightjar, Hobby, Garganey, and Stone-Curlew. In some instances, however, it is possible that certain of the species named may have passed the winter in our islands, especially in the southern counties; indeed a few or single individuals of the Ring-Ouzel, Blackcap, Chiff-Chaff, Garganey, and Stone-Curlew have been known to do so in seasons of unusual mildness.

Emigration of Winter Visitors.—In genial seasons, Skylarks and Blackbirds which have been winter guests

---

[1] For particulars of dates of arrival of Summer Visitors, see pp. 126-128.

in our islands are sometimes observed from the third week of February at the lighthouses, and other suitably situated watching stations, under circumstances which leave little doubt that they were departing for their northern native lands. "Grey Geese," Brent Geese, and Iceland Gulls have also been observed moving northwards at the same period.

The chief emigratory movements of the month, however, are those of Rooks and Skylarks departing across the southern waters of the North Sea en route for Western Central Europe, whence they came in the autumn to pass the winter with us. These emigrations are embarked upon on the south-east coast of England, and the movements are much in evidence at the lightships stationed off the coasts from Norfolk to Kent, where the birds are observed, during the daytime, proceeding towards the east and south-east, *i.e.*, in the direction of Holland, Belgium, and north-eastern France.

MARCH.—In its climatic aspects March is a variable month. The advance in temperature over that of February is comparatively small in the British Isles, though on the Continent the increase is rapidly proceeding. With us the average amount of sunshine is 7 per cent. of the annual, as against 5 per cent. for the previous month. Though March witnesses a conflict between expiring winter and the advance of spring, yet much migration of a varied nature is performed, and hence it is an important month in the calendar of the bird-watcher.

Local Movements.—The local migrations from British winter to summer haunts are much in evidence, and relate (in addition to the birds already mentioned for

February) to the return to their nesting quarters of the Pied and Grey Wagtails, Twite, Merlin, Mallard, Teal, Woodcock, Black-headed Gull, and others ; while many Ring-Plovers, Richardson's Skuas, and Fulmar Petrels appear for the summer in the northern archipelagos of Orkney and Shetland, and other boreal localities. The local movements also include the departure from the islands off the west coasts of Scotland and Ireland of a number of species which, having passed the winter in the milder climes of the Far West, are returning to their summer quarters in Great Britain or the Sister Isle. Such migrants consist chiefly of Greenfinches, Chaffinches, Twites, Song-Thrushes, Blackbirds, and Starlings.

Arrival of Partial Migrants.—The return journeys from their southern Continental winter retreats of the Partial Migrants are continued, and reach their maximum during March, all the species recorded as arriving on the south coast of England for February being observed, and, in addition, the Woodcock and Curlew. Considerable numbers of Rooks (occasionally accompanied by Jack-daws) and of Starlings and Skylarks arrive during the daytime on the south-east coast between Kent and Norfolk throughout the month, and their immigrations are sometimes prolonged into the first week of April. These arrivals sometimes occur for several successive days, and the birds returning are doubtless those which were observed leaving these same shores for those of the Continent during the previous autumn.

Arrival of Summer Visitors.[1]—The Summer Visitors which have been known to appear in our islands during March are thirty-eight in number. Of these, however,

[1] For particulars of dates of arrival of Summer Visitors, see pp. 126-128.

only about ten can be regarded as being regular in
their appearance, chiefly in the southern counties of
England, and most of them during the latter half of the
month. These regular March summer birds are the
Ring-Ouzel, Wheatear, Chiff-Chaff, Willow-Warbler,
Swallow, Sand-Martin, Wryneck, Garganey, and Sandwich Tern. Others less constant, but not infrequent in
their visits, are the Redstart, Blackcap, White Wagtail
(for passage), Yellow Wagtail, House - Martin, and
Common Sandpiper :[1] while the recorded appearance of
the Whinchat, Nightingale, Common and Lesser Whitethroats, Garden-Warbler, Sedge-Warbler, Tree-Pipit,
Spotted and Pied Flycatchers, Red-backed Shrike, Swift,
Cuckoo, Hoopoe, Hobby, Corn - Crake, Dotterel, Ruff
and Reeve, Whimbrel, and Lesser Tern, must be
regarded as quite exceptional.

The following summer birds reach Scotland during
the month, with more or less regularity :—The Ring-
Ouzel, Wheatear, Chiff - Chaff, Sand - Martin, and
Swallow ; rarely the House - Martin ; very rarely the
Willow-Warbler and the Redstart. The Lesser Black-
backed Gull, which is chiefly a summer visitor in
Scotland, also appears during the month.

The Ring-Ouzel, Wheatear, Chiff-Chaff, and Sand-
Martin are regular March visitors to Ireland. The
Swallow is seldom observed before April, but appeared
over a wide area during the latter half of March in 1903.

The meteorological character of the month has a
marked effect on the date of appearance of these first
arrivals among the summer birds. Thus during the
prolonged spell of genial weather which rendered March

---

[1] For the date of first appearance, and usual dates of arrival of the
Summer Visitors, see pp. 126-128.

in 1884 remarkable for a warmth exceeding that experienced during this month for many years, no less than sixteen species of spring migrants were recorded as having appeared in our islands. The March of 1886 was another month of exceptional geniality, and it, too, had a goodly show of spring birds. On the other hand, the March of 1883 was cold in the extreme, and only the Ring-Ouzel, Wheatear, Chiff-Chaff, and Swallow were noted. In 1885 the month was also remarkable for its ungeniality, and likewise for the fewness of its spring birds.

Along with the British summer guests there arrive on the south coast numerous Redwings and Fieldfares from the south. These birds are returning to their summer homes in Northern Europe, but it is doubtful if they proceed beyond our shores during the month.

Emigration of Winter Visitors.—March witnesses the beginning of the great departure movements, for their summer haunts, of the birds which have been winter guests in our islands. The migrations northwards are much in evidence at suitably situated stations on and near the more boreal sections of our coast-line ; and much information concerning the species participating in them, and the dates at which they are performed, will be afforded in the special study devoted to Fair Isle. The departing visitors quit the mainland during the night, and many make this, and other favourably situated isles of the Orkney and Shetland group, a resting place.

The species quitting the British Isles and moving northwards during March include Song - Thrushes, Blackbirds, Redbreasts, Goldcrests, Chaffinches, Bramblings, Mealy Redpolls, Yellow Buntings, Reed Bunt-

I.   H

ings, Lapland Buntings, Starlings, Rooks, Skylarks, Shore Larks, Short-eared Owls, "Grey Geese," Brent Geese, Whooper and Bewick's Swans, Mallards, Gadwalls, Wigeons, Golden-eyes, Scaup, Long-tailed Ducks, Red-breasted Mergansers, Golden Plovers, Lapwings, Woodcocks, Dunlins, Curlews, Little Auks, and Redthroated Divers.

The emigratory movements from the south-east coast of England in an easterly direction towards the coasts of Holland, Belgium, and north-eastern France, which commenced during February, become more pronounced, and the travellers include Black Redstarts, Tree-Sparrows, Chaffinches, Starlings, Hooded Crows, Rooks, Jackdaws, and Skylarks. Numbers of some of these species, the Grey Crows in particular, move southwards along the shores of Yorkshire and Lincolnshire in order to reach the scene of embarkation on the coasts of Norfolk, Suffolk, and Essex.

APRIL.—This is the first month of the year in which the increase of temperature usually makes itself pronouncedly felt over both our insular and the continental areas. In the British Isles the isotherms then make the nearest approach to straight lines, though with a slant from W.N.W. to E.S.E. It is a month of much sunshine, the amount rising to 13 per cent. of the annual. The winter type of distribution of temperature, in which the inland values are lower than those on the coast, disappears, and the summer type, with its high inland values and low coast values, takes its place, and prevails until October.

Throughout the month the movements of migrants are of a pronounced nature.

Local Movements. — The following species, which have wintered on our shores or their neighbourhood, return to their nesting haunts during the month :— The Dunlin, Fork-tailed Petrel, Shelduck, Eider Duck, Great Skua, Arctic Skua, Puffin, and Razorbill; in addition to others which commenced their return in March.

Arrival of Summer Visitors.[1]—All the species which are summer visitors to our islands (see page 46), with the exception of the Marsh-Warbler and the Red-necked Phalarope, arrive on our shores. The inflowing stream is a continuous one during the prevalence of meteorological conditions suited for their journeys ; but it is not until the second week of the month that the main body of these summer guests appears, and the flowing tide is maintained to and beyond the close of the month. Some species, however, are decidedly irregular in their appearance thus early in the season, among them the Nightjar ; while others, such as the Whinchat, Common and Lesser Whitethroats, Garden-Warbler, Wood-Warbler, Reed-Warbler, Sedge-Warbler, Red-backed Shrike, Pied and Spotted Flycatchers, Swift, Turtle Dove, Land-Rail, Dotterel, Arctic, Roseate, and Little Terns arrive more abundantly in May.

Immediately after arrival on the southern shores of England, our summer birds proceed north, east, and west to reach their accustomed nesting haunts in various parts of the British area.

Some of these April immigrations not only comprise a great multitude of individuals, but include many species moving in company. Thus at the Eddystone

[1] For particulars of dates of earliest appearance, and usual dates of arrival of the Summer Visitors, see pp. 126-128.

lighthouse, on the night of 11th April and the early morning of the 12th, 1902, a vast movement was witnessed, the weather at the time being suitable for bringing the birds under notice of the observers—namely, a moderate breeze accompanied by thick rain, which prevailed the whole of the night. The migration first became apparent at 8.45 P.M., with the arrival of Wheatears, which passed continuously to the north-west in great numbers until 11 P.M. These were closely followed by numerous Song-Thrushes ; and at midnight hundreds of Ring-Ouzels and Redwings came upon the scene, accompanied by Starlings, Swallows, and "a perfect cloud of small birds," composed of Redstarts, Nightingales, Blackcaps, Tree-Pipits, and many other species, examples of which were either not killed at the lantern, or, being injured, were lost in the sea below. Great numbers of Wheatears again appeared at 1.30 A.M., and continued, with the other species, to pass northwards until 5 A.M. From midnight until 4 A.M. "the air seemed to be thick with birds, but they melted away, as it were, on the appearance of dawn." Examples of all the species named were submitted to me for identification. At the Owers light-vessel, stationed off the east side of the Isle of Wight, many Cuckoos, Redwings, and a great number of "warblers," occurred at the same time, and many were killed by striking against the lantern.

These records from the Eddystone and the Owers are of further interest, since they clearly indicate that in addition to the summer guests making for our islands, such as the Nightingale, Tree-Pipit, Cuckoo, etc., etc., there were also simultaneously seeking our shores a number of birds on passage to countries beyond our area—namely, Redwings, Song-Thrushes, and Starlings.

The Redwings were obviously on their way to their northern breeding grounds, and there can be little doubt that the Song-Thrushes and Starlings arriving at that date were on a similar errand. The mixture of migrants indicated is not the exception, but the rule, at this period of the season ; and later it becomes quite impossible to discriminate, in the case of identical species, between individuals which are bent upon spending the summer in our isles, and those which are en route to nesting haunts beyond our shores. These remarks do not apply to the Nightingale, Reed-Warbler, Marsh-Warbler, Grasshopper Warbler, Thicknee or Stone Curlew, Kentish Plover, etc., which are summer visitors only, and do not occur as birds of passage on our shores.

Passage Movements.[1]—The arrivals on our shores during April include a fresh set of migrants for the season—namely, the Birds of Passage which are en route from their southern winter quarters to their summer homes to the north and east of us.

The first of these arrivals appear in company with our summer visitors, as has just been related, and they often leave us in company with birds of the same species which have spent the winter in our midst, such as the Redwing, Fieldfare, and many others, which are likewise proceeding to their northern summer quarters. When this is the case, it is impossible to distinguish between the individuals on passage and emigrating British winter visitors ; the birds of passage, however, seldom make their way inland, but traverse the coast-lines and their vicinity, though they be Thrushes, Warblers, or other land-birds.

---

[1] For a list of the Birds of Passage, with the dates between which their movements are performed, see page 129.

The Birds of Passage which are observed in our islands during the month number about seventy species, and include (apart from species which are also British emigrants) the Ring-Ouzel, Blackcap, Willow-Warbler, Redstart, White Wagtail, Whinchat, Ortolan Bunting, Pied Flycatcher, Swallow, Sand-Martin, Wryneck, Green Sandpiper, Wood Sandpiper, Little Stint and Whimbrel. The majority of these occur during the latter half of the month, and are on their way to northern lands, but some of them are not observed every year as passing April migrants.

Emigration of Winter Visitors.—The departure of those birds which have spent the winter in our islands sets in in earnest during April. Among the emigrants proceeding to their northern homes are the Goldcrest, Hedge Accentor, Great Grey Shrike, Siskin, Mealy Redpoll, Short-eared Owl, Kestrel, Grey Lag, White fronted, Pink-footed, Barnacle, and Brent Geese, Whooper and Bewick's Swans; among ducks the Tufted, Golden-eye, and Long-tailed species; the Water-Rail, Woodcock, Jack Snipe, Greenshank, and Glaucous Gull; and among rarer species the Little and Lapland Buntings, Snowy Owl, Rough-legged Buzzard, Greenland Falcon, and Smew.

The departure movements of the Central European birds among these winter guests, which commenced in February, are still in progress from the south-east coast of England, but terminate with the month. These emigrants to the eastwards are of the same species as those recorded for the previous month, but it is probable that the Hoopoes which traverse the south coast at this season may also find their way to their Continental summer homes by this North Sea route.

MAY.—In May the tide of spring migration rises to its maximum height.   The arrivals in and departures from our islands are many, and the numbers of the migrants great, but the most important, so far as volume is concerned, are the movements of birds on their passage from the south to the north, which are in progress along our shores throughout the month.

Arrival of Summer Visitors.[1]—Our summer guests still pour into Great Britain and Ireland, especially during the first half of the month.   This is particularly the case with those species whose average date of first arrival is in the latter half of April, such as the Lesser Whitethroat, Garden-Warbler, Grasshopper Warbler, Reed-Warbler, Red-backed Shrike, Spotted Flycatcher, Pied Flycatcher, Nightjar, Swift, Honey-Buzzard, Turtle Dove, Corn-Crake, Quail, Dotterel, Common Tern, Arctic Tern, Lesser Tern, and Red-necked Phalarope, all of which are characteristic of May in most parts of the British area. On the other hand, certain species which are among the earliest summer birds to arrive, such as the Ring-Ouzel, Wheatear, Chiff-Chaff, Swallow, Sand-Martin, Yellow Wagtail, Willow-Warbler, Blackcap, Wryneck, and Sandpiper, do not appear as summer residents beyond the earliest days of the month, though they arrive on and traverse our shores as birds of passage throughout its days.

Sometimes we have a sharp and melancholy reminder that

"Winter, lingering, chills the lap of May."

This was the case in 1886, when in the second week of May, after several very cold days for the time of the

---

[1] For particulars of dates of arrival of the Summer Visitors, see p. 126.

year, a severe north-east gale set in, accompanied by heavy rain and sleet in the valleys, and several inches of snow on the fells. During this period the insect-feeding birds perished in thousands, Swallows, House-Martins, Sand-Martins, and Swifts being the greatest sufferers.

Passage Movements.[1]—May, as we have said, is pre-eminently the month for the passage northwards along our shores of the vast numbers of birds which, having passed the winter in southern regions, are on their way to spend the summer in the sub-arctic and arctic lands from Greenland to Western Siberia. An important climatic feature of the month, and one that has con-siderable bearings upon these movements, is the rise of temperature that takes place in the inland portions of Scandinavia, rendering this great peninsula suitable for the return of its vast array of summer birds.

These birds arrive on our southern shores and pro-ceed northwards with little or no delay, travelling along both the east and west coast routes. On the advent of suitable weather-conditions following decidedly adverse periods, the movements take the form of impetuous rushes, which are participated in by species representing widely different orders, from delicate Warblers, Wag-tails, Finches, Swallows, etc., to Birds of Prey, Plovers, Sandpipers, Geese, Skuas, etc. Pronounced movements are, indeed, not uncommon during May, and the migrations of the great majority of the species cover the entire month.

The May birds of passage, not including those species which are also emigrating British winter visitors, to be mentioned immediately, are the Ring-Ouzel, Wheatear,

[1] For particulars of the dates of Passage Movements, see p. 129.

Greater Wheatear, Whinchat, Redstart, Whitethroat, Lesser Whitethroat, Blackcap, Garden-Warbler, Willow-Warbler, Chiff-Chaff, Sedge-Warbler, White Wagtail, Grey-headed Wagtail, Tree-Pipit, Red-backed Shrike, Pied and Common Flycatchers, Swallow, Martin, Ortolan Bunting, Swift, Nightjar, Wryneck, Cuckoo, Honey-Buzzard, Osprey, Corn-Crake, Dotterel, Red-necked Phalarope, Great Snipe, Little Stint, Curlew Sandpiper, Common Sandpiper, Wood Sandpiper, Green Sandpiper, Spotted Redshank, Black-tailed Godwit, Whimbrel, and Buffon's Skua.

Much information on these movements will be afforded in the studies devoted to Fair Isle.

Emigration of Winter Visitors.—May, especially its first half, is also a month for much emigration on the part of those northern birds which have wintered in our islands. These departure movements, as already stated, are much mixed with those of the birds of passage also proceeding northwards at the same time; and when the species are identical, it is impossible to discriminate between them. These birds flit away from us during the night, taking their departure from the eastern and northern coasts.

These May emigrants (those marked thus * being also birds of passage) are the *Mistle-Thrush, *Field-fare, *Redwing, *Blackbird, *Song-Thrush, *Redbreast, *Hedge-Accentor, *Meadow-Pipit, *Chaffinch, *Bram-bling, *Siskin, Mealy Redpoll, Yellow Bunting, Reed-Bunting, *Snow Bunting, *Short-eared Owl, *Long-eared Owl, *Ring Dove, *White-fronted Goose, Pink-footed Goose, Barnacle Goose, *Wigeon, *Pintail, Scaup, Golden-eye, Long-tailed Duck, *Merganser, Ringed Plover, Lapwing, *Golden Plover, *Grey Plover,

*Woodcock, *Turnstone, *Grey Phalarope, *Snipe, *Jack Snipe, *Dunlin, Purple Sandpiper, *Knot, *Redshank, *Sanderling, *Bar-tailed Godwit, *Curlew, Little Grebe, *Great Northern Diver, and *Black-throated Diver.

JUNE. — June witnesses the close of the spring migratory movements. Few birds arrive as summer visitors during the month, but the Marsh-Warbler is one of them, being the latest of all to appear.

Several species of Birds of Passage whose summer homes lie within the arctic circle, or amid the polar regions, continue to traverse our shores, some of them in considerable numbers, during the earlier days of the month, and even beyond that period.

Among these latest spring travellers are the Mealy Redpoll, Snow-Bunting, Greater Wheatear, Osprey, "Swans," "Wild Geese," Golden-eye, Scaup, Golden Plover, Grey Plover, Knot, Turnstone, Red-necked Phalarope, Purple Sandpiper, Dunlin, Little Stint, Sanderling, Bar-tailed Godwit, Black-tailed Godwit, Whimbrel, Pomatorhine Skua, Buffon's Skua, Great Northern Diver, and Red-throated Diver. Others there are which have only occasionally been recorded for the month, but whose appearance may not, perhaps, be so unusual as the scanty data would lead us to suppose. These are the Honey - Buzzard, Barnacle Goose, Wigeon, Long-tailed Duck, Jack Snipe, Ruff, Green Sandpiper, Wood Sandpiper, Spotted Redshank, and Greenshank; while Black Terns and Spoonbills have been noted as visitors to the south-east coast of England.

In addition to these, individuals of a number of

species appear at stations where there is no mistaking the nature of their visits. These are the remnants of the rear-guard of the feathered army which has already passed to its summer quarters, and in its ranks are to be found Swallows, Martins, Red-backed Shrikes, Tree-Pipits, Spotted Flycatchers, Willow-Warblers, Garden-Warblers, Blackcaps, Common Whitethroats, Lesser Whitethroats, Redstarts, Arctic Bluethroats, Swifts, Nightjars, Cuckoos, Wrynecks, Corn-Crakes, and Common Sandpipers.

In connection with the arrival of the Summer Visitors, an interesting fact remains to be related—one that was first made known through the " Digest of Observations " submitted to the British Association in 1896. When studying the vast data amassed by the Migration Committee, I found that the great majority of our summer birds appeared on the west coast of England some days in advance of their arrival in their eastern haunts. More recent investigations have confirmed this, and have made known to us that the same rule applies to Scotland, where the Solway and Clyde areas receive their first spring migrants some days earlier than the areas of Tweed and Forth. Thus the Swallow arrives on the south-west coast of Scotland several days earlier than in the south-east. It is not unnatural that the British summer birds should seek first those portions of our islands which are the most genial at this early period of the season. A reference to a map giving the spring isotherms, shows us that in March the Solway has an average temperature equal to that of the Thames, the Clyde to that of the Tyne ; and that in April it is as warm in Cantyre as it is about the Wash and Humber.

Another feature of the spring migration, already briefly alluded to, is worthy of consideration. We have seen that the Birds of Passage are the latest of all the spring migrants to appear on our shores. At the time that many of them are traversing our islands on their way to the north, the British residents and summer visitors of the same species are already either deeply occupied in the incubation of their eggs or actually tending their broods—among others the Song-Thrush, Blackbird, Ring - Ouzel, Wheatear, Golden Plover, Woodcock, Dunlin, Snipe, Redshank, Common Sandpiper, and Curlew.

In most cases it would be utter folly for the migrants bound for the Far North to seek their summer haunts there at earlier periods, since the climatic conditions then prevailing render them totally unsuited for their reception.[1]

Spring Casual Visitors.—A considerable number of rare visitors of various species appear annually in the British Islands as waifs during the period covered by the spring migratory movements. Each year produces a more or less abundant crop. Most of them belong to species whose breeding haunts are in Southern and Central Europe, and some may have reached our shores by accident, when journeying from their African winter quarters to their native land, having for some reason overshot their customary northern or western limit. These visits may be, and no doubt often are, accounted for by the voyagers having encountered strong southerly or easterly winds when en route for their Continental summer haunts, and thus having been swept out of their

[1] See Plate III., showing the advance of spring.

course reach our shores. Good examples of migrants hopelessly lost at this season are furnished by the occurrence of such species as the Desert Wheatear at the Pentland Skerries, Savi's Warbler at Fair Isle, the Subalpine Warbler at St Kilda and Fair Isle, the Red-rumped Swallow at Fair Isle, and many other instances equally astonishing of the appearance in our islands of waifs and wanderers from far-off lands.

Other casual visitors at this season belong to species in which the nomadic habit is strongly developed, such as the Rose-coloured Pastor or Starling, and Pallas' Sand-Grouse. The visits of such species are, of course, more or less erratic, and the birds appear at irregular intervals and in varying numbers.

[APPENDIX I.

APPENDIX I.—DATES OF THE ARRIVAL OF SUMMER VISITORS

THE information relating to the appearance of the harbingers of spring is very voluminous, and more complete than that relating to the movements of any other set of migrants.

Species	Early Records.	Usual Date of 1st Arrival, England.	Usual Date of 1st Arrival, Scotland.	Period of Arrival in Numbers.[1]
GOLDEN ORIOLE	15th April 1890	28th April	.	.
WHITE WAGTAIL	3rd March 1872	20th March	8th April	27th March to 14th May.
YELLOW WAGTAIL	10th March 1878	22nd March	15th April	8th April to 11th May.
BLUE-HEADED WAGTAIL	.	25th April	.	.
TREE-PIPIT	22nd March 1893	4th April	20th April	7th April to 17th May.
RED-BACKED SHRIKE	21st March 1880	22nd April	.	4th to 25th May.
WHITETHROAT	10th March 1881	8th April	26th April	12th April to 28th May.
LESSER WHITETHROAT	18th March 1881	11th April	.	19th April to 22nd May.
BLACKCAP	3rd March 1871	22nd March	20th April	4th April to 23rd May.
GARDEN-WARBLER	28th March 1879	15th April	4th May	22nd April to 28th May.
WOOD-WARBLER	16th March 1881	12th April	25th April	18th April to 21st May.

WILLOW-WARBLER	9th March 1871	23rd March	9th April	29th March to 26th May.
CHIFFCHAFF	24th February 1873	16th March	2nd April	20th March to 3rd May.
SEDGE-WARBLER	22nd March 1871	6th April	24th April	18th April to 28th May.
REED-WARBLER	5th April 1909	11th April	·	19th April to 28th May.
MARSH-WARBLER	30th May	Early June.	·	·
GRASSHOPPER-WARBLER	2nd April 1908	18th April	30th April	20th April to 13th May.
RING-OUZEL	17th February	12th March	5th April	25th March to 11th May.
NIGHTINGALE	22nd March 1893	8th April	·	13th April to 9th May.
REDSTART	23rd March 1864	8th April	13th April	11th April to 10th May.
WHEATEAR	24th February 1885	12th March	18th March	16th March to 24th May.
WHINCHAT	9th March 1884	8th April	18th April	17th April to 22nd May.
SPOTTED FLYCATCHER	10th March 1882	18th April	4th May	28th April to 28th May.
PIED FLYCATCHER	26th March 1878	19th April	·	25th April to 15th May.
SWALLOW	24th February 1886	21st March	6th April	31st March to 15th May.
HOUSE-MARTIN	9th March 1871	6th April	17th April	10th April to 28th May.
SAND-MARTIN	29th February 1886	19th March	4th April	25th March to 22nd May.
WRYNECK	12th March 1884	21st March	·	1st April to 9th May.
SWIFT	21st March 1892	24th April	2nd May	27th April to 28th May.
NIGHTJAR	4th April 1869	18th April	26th April	28th April to 25th May.
CUCKOO	10th March 1884	3rd April	18th April	9th April to 22nd May.

¹ The latest dates quoted for certain species may relate to arrivals of birds on passage to northern Europe.

DATES OF THE ARRIVAL OF SUMMER VISITORS—*continued*

Species.	Early Records.	Usual Date of 1st Arrival, England.	Usual Date of 1st Arrival, Scotland.	Period of Arrival in Numbers.
HOBBY	3rd March 1866	19th April	.	.
GARGANEY	24th February 1883	19th March	.	.
TURTLE-DOVE	4th April 1909	23rd April	.	27th April to 26th May.
QUAIL	20th April 1904	29th April	.	16th May.
CORN-CRAKE	28th March 1884	17th April	22nd April	20th April to 15th May.
STONE-CURLEW	14th March 1904	21st March	.	9th April.
DOTTEREL	25th March 1843	20th April	.	to 26th May.
KENTISH PLOVER	24th March 1909	17th April	.	to 21st April.
RED-NECKED PHALAROPE	16th May	. .	21st May	.
COMMON SANDPIPER	16th March 1896	24th March	11th April	8th April to 23rd May.
WHIMBREL	24th March 1887	18th April	26th April	26th April to 9th June.
SANDWICH TERN	23rd March 1887	29th March	18th April	26th April.
COMMON TERN	9th March 1850	16th April	27th April	16th April to 20th May.
ARCTIC TERN	18th April 1866	28th April	14th May	24th May.
ROSEATE TERN	30th April 1897	May	.	.
LITTLE TERN	7th March 1850	16th April	29th April	16th April to 17th May.

APPENDIX II.—Dates of the Movements of the Birds of Passage in Spring and Autumn

The following data have been mainly derived from observations made at stations where the various species are visitors on passage only.

Much remains to be ascertained regarding the times of the migrations of certain of the Ducks, Geese, Gulls, Divers, and other swimming birds, whose passages are performed at sea, and hence largely escape notice. The periods of the passage movements of some of these, and of several other species, are not given, as the data available are too fragmentary to be of value. The latest dates quoted often, no doubt, relate to the appearance of stragglers :—

Grey Crow.

Spring.—9th March to 12th May.
Autumn.—3rd August to 2nd November. Chiefly October..

Starling.

Spring.—22nd March to 17th April.
Autumn.—9th September to 22nd November. Chiefly October.

Chaffinch.

Spring.—22nd March to 12th May. Chiefly April.
Autumn.—3rd September to 20th November. Chiefly October to mid-November.

Brambling.

Spring.—8th April to 20th May.
Autumn.—17th September to 15th November. Chiefly late September and in October.

I.   I

SISKIN.

Spring.—19th April to 13th May.
Autumn.—22nd September to 25th November.

MEALY REDPOLL.

Spring.—14th March to 29th May.
Autumn.—21st September to 28th November. Chiefly latter half of October.

REED-BUNTING.

Spring.—26th March to 28th May. Chiefly second week of April to mid-May.
Autumn.—20th September to 11th November.

YELLOW BUNTING.

Spring.—22nd March to 23rd May.
Autumn.—17th September to 19th November.

ORTOLAN BUNTING.

Spring.—30th April to 6th June. Chiefly mid-May.
Autumn.—29th August to 16th October. Chiefly September.

SNOW-BUNTING.

Spring.—12th March to 8th June. Chiefly latter half of March and early April.
Autumn.—8th September to 10th November. Chiefly October.

LAPLAND BUNTING.

Spring.—25th March to 2nd May.
Autumn.—25th August to 29th October. Chiefly September and early October.

SHORE-LARK.

Spring.—18th March to 22nd April.
Autumn.—11th September to 17th November.

SKYLARK.

Spring.—March to 7th May to north. Mid-February to end of March to east.
Autumn.—17th September to 15th November. Chiefly October.

PIED WAGTAIL.

Spring.—9th March to 8th May.   Chiefly late March to mid-April.
Autumn.—22nd July to 4th November.   Chiefly August.

WHITE WAGTAIL.

Spring.—7th April to 31st May.   Chiefly 1st to 21st May.
Autumn.—9th August to 17th November.   Chiefly September.

GREY-HEADED WAGTAIL.

Spring.— 5th May to 3rd June.   Chiefly latter half of May.
Autumn.—26th September to 4th November.

TREE-PIPIT.

Spring.—5th May to 10th June.   Chiefly latter half of May.
Autumn.—26th August to 9th November. Chiefly latter half of September.

MEADOW-PIPIT.

Spring.—20th March to 22nd May.   Chiefly late April to mid-May.
Autumn.—29th August to 18th November.   Chiefly September to mid-October.

GOLDCREST (Continental race).

Spring.—25th March to 8th May.   Chiefly April.
Autumn.—8th September to 25th November.   Chiefly late September and during October.

GREAT GREY SHRIKE.

Spring.—6th April to 27th April.
Autumn.—1st September to 30th November.

RED-BACKED SHRIKE.

Spring.—5th May to 4th June.
Autumn.—5th August to 3rd October.   Chiefly September.

WHITETHROAT.

Spring.—4th May to 8th June.
Autumn.—27th August to 9th November.   Chiefly September.

LESSER WHITETHROAT.

Spring.—23rd April to 9th June.   Chiefly last three weeks of May.
Autumn.—20th August to 6th November.   Chiefly September.

BLACKCAP.

Spring.—28th April to 3rd and 15th June.   Chiefly latter half of May.
Autumn.—24th August to 22nd November.   Chiefly September.

GARDEN-WARBLER.

Spring.—9th May to 10th June.   Chiefly mid-May.
Autumn.—17th August to 25th November.   Chiefly September.

BARRED WARBLER.

Autumn.—3rd August to 13th November.   Chiefly September.

WILLOW-WARBLER.

Spring.—15th April to 6th June.   Chiefly first three weeks of May.
Autumn.—7th August to 13th November.   Chiefly September.

CHIFFCHAFF, *P. collybita abietina.*

Spring.—7th May to 1st June.   Chiefly middle weeks of May.
Autumn.—26th September to 18th November.   Chiefly late September
and early October.

YELLOW-BROWED WARBLER.

Spring.—11th April, 1909.
Autumn.—16th September to 29th October.   Chiefly late September
to mid-October.

ICTERINE WARBLER.

Spring.—4th May to 1st June.
Autumn.—4th to 29th September.

SEDGE-WARBLER.

Spring.—7th May to 3rd June.   Chiefly middle weeks of May.
Autumn.—10th August to 29th November.   Chiefly September.

MISTLE-THRUSH.

Spring.—6th March to 6th May.
Autumn.—1st October to 20th November.   Chiefly late October.

SONG-THRUSH.

Spring.—22nd March 1909, and 8th April to 22nd May.   Chiefly late
April to early May.
Autumn.—4th September to 25th November.   Chiefly latter half of
October to third week of November.

REDWING.

Spring.—22nd March 1909, and 8th April to 12th May.   Chiefly April.

Autumn.—19th September to 22nd November.   Chiefly October and November.

FIELDFARE.

Spring.—22nd March 1909, and 8th April to 28th May.   Chiefly second week of April to mid-May.

Autumn.—22nd September to 25th November.   Chiefly latter half of October to end of third week of November.

BLACKBIRD.

Spring.—24th March to 12th May.   Chiefly April.

Autumn.—14th September to 29th November.   Chiefly after mid-October.

RING-OUZEL.

Spring.—18th March 1906, and 8th April to 21st May.   Chiefly last week April to mid-May.

Autumn.—24th August, and 23rd September to 27th November.   Chiefly latter half of October.

REDBREAST (Continental race).

Spring.—8th April to 23rd May.   Chiefly April and first week of May.

Autumn.—20th September to 24th November.   Chiefly latter half of October and first week of November.

RED-SPOTTED BLUETHROAT.

Spring.—24th April to 2nd June.   Chiefly mid-May.

Autumn.—3rd September to 11th October.   Chiefly latter half of September.

REDSTART.

Spring.—22nd March 1909, and 4th April to 6th June.   Chiefly second and third weeks of May.

Autumn.—29th August to 7th November.   Chiefly latter half of September.

WHEATEAR.

Spring.—5th April to 12th May.   Chiefly late April.

Autumn.—15th August to 14th November.   Chiefly September.

GREATER WHEATEAR.

Spring.—27th April to 1st June.
Autumn.—5th September to 20th November.   Chiefly September.

WHINCHAT.

Spring.—30th April to 6th June.   Chiefly second and third weeks of May.
Autumn.—24th August to 9th November.   Chiefly September.

SPOTTED FLYCATCHER.

Spring.—12th May to 14th June.
Autumn.—16th August to 30th November.   Chiefly September.

PIED FLYCATCHER.

Spring.—15th and 30th April to 24th May.   Chiefly middle weeks of May.
Autumn.—24th August to 5th November.   Chiefly September.

SWALLOW.

Spring.—17th and 30th April to 8th June.   Chiefly second week to end of May.
Autumn.—9th September to 13th November.   Chiefly mid-September to mid-October.

MARTIN.

Spring.—7th May to 8th June.   Chiefly late May and early June.
Autumn.—September to 16th November.   Chiefly mid-September to mid-October.

SAND-MARTIN.

Spring.—28th April to 1st June.
Autumn.—September to 12th November.

SWIFT.

Spring.—8th May to 23rd June.   Chiefly late May.
Autumn.—July to 16th November.   Chiefly August.

NIGHTJAR.

Spring.—20th May to 10th June.
Autumn.—17th August to 18th October.

WRYNECK.

Spring.—30th April to 4th June. Chiefly second and third weeks of May.
Autumn.—19th August to 16th November. Chiefly September.

HOOPOE.

Spring.—30th March to 20th May.
Autumn.—8th August to 21st November.

CUCKOO.

Spring.—8th May to 8th June. Chiefly second and third weeks of May.
Autumn.—16th August to 26th November. Chiefly latter half of August.

LONG-EARED OWL.

Spring.—14th March to 23rd May.
Autumn.—13th October to 16th November.

SHORT-EARED OWL.

Spring.—25th March to 12th May. Chiefly April and early May.
Autumn.—28th September to 13th November. Chiefly latter half of October.

HONEY-BUZZARD.

Spring.—21st May to 11th June.
Autumn.—29th July to 22nd October. Chiefly September.

MERLIN.

Autumn.—30th August to 12th November.

KESTREL.

Spring.—16th March to 18th May.
Autumn.—25th August to 12th November. Chiefly September.

OSPREY.

Spring.—May and early June.
Autumn.—September and October to 17th November. Chiefly September and October.

MALLARD.

Spring.—25th March to 14th May.
Autumn.—12th September to 28th November.

TEAL.

Spring.—25th March to 11th May.
Autumn.—30th August to 29th November.

WIGEON.

Spring.—25th March to 11th May and 12th June.
Autumn.—30th August to 23rd November.   Chiefly October.

TUFTED DUCK.

Spring.—24th April to 10th May.
Autumn.—29th October to 27th November.

SCAUP-DUCK.

Spring.—24th March to 12th June.
Autumn.—22nd September to 9th November.

GOLDENEYE.

Spring.—23rd March to 12th June.
Autumn.—14th September to 21st November.

COMMON SCOTER.

Spring.—29th April to 7th June.
Autumn.—5th September to late October.

VELVET-SCOTER.

Spring.—31st April and during May.
Autumn.—17th September to 20th October.

RED-BREASTED MERGANSER.

Spring.—16th March to 14th May.   Chiefly early May.
Autumn.—3rd September to 31st October.   Chiefly September and
early October.

RING-DOVE.

Spring.—23rd March to 31st May.   Chiefly April and May.
Autumn.—25th September to 10th November.   Chiefly mid-October.

CORN-CRAKE.

Spring.—30th April to 2nd June.
Autumn.—20th August to 19th November.

WATER-RAIL.

Spring.—29th March to 4th May.
Autumn.—21st September to 17th November.

DOTTEREL.

Spring.—5th May to 15th June.
Autumn.—16th August to 26th October.

RINGED PLOVER.

Spring.—23rd March to 16th May.
Autumn.—2nd September to 14th November.

GOLDEN PLOVER.

Spring.—22nd March to 30th May.   Chiefly mid-April to mid-May.
Autumn.—2nd September to 30th November.   Chiefly latter half of September.

GREY PLOVER.

Spring.—30th March to 7th June.   Chiefly May.
Autumn.—7th August to 9th November.   Chiefly September.

LAPWING.

Spring.—22nd March to 21st May.   Chiefly April.
Autumn.—7th September to 18th November.   Chiefly October.

TURNSTONE.

Spring.—30th March to 4th June.   Chiefly May.
Autumn.—14th August to 21st November.   Chiefly September.

OYSTER-CATCHER.

Spring.—16th March to 2nd May.
Autumn.—September to 14th November.

RED-NECKED PHALAROPE.

Spring.—16th May to 19th June.
Autumn.—September to 11th November.

WOODCOCK.

Spring.—23rd March to 7th May.
Autumn.—23rd September to 22nd November.   Chiefly fourth week
of October and first week of November.

GREAT SNIPE.

Spring.—5th to 15th May.
Autumn.—26th July to 16th November.   Chiefly September and
October.

COMMON SNIPE.

Spring.—9th April to 9th May.
Autumn.—4th September to 26th November.

JACK SNIPE.

Spring.—24th March 1909, and 16th April to 20th May.
Autumn.—11th September (20th August 1893) to 24th November.
Chiefly latter half of September and early October.

DUNLIN.

Spring.—22nd March to 9th June.   Chiefly May.
Autumn.—9th August to 4th November.   Chiefly late August and
during September.

LITTLE STINT.

Spring.—20th April to 18th June.
Autumn.—27th July to 9th November.   Chiefly September.

CURLEW SANDPIPER.

Spring.—27th April to 31st May.
Autumn.—20th July to 18th November.   Chiefly late August and early
September.

PURPLE SANDPIPER.

Spring.—1st April to 5th June.
Autumn.—4th August to 17th October.

KNOT.

Spring.—6th April to 28th May.
Autumn.—13th August to 30th November. Chiefly latter half of August and during September.

SANDERLING.

Spring.—6th April to 12th June.
Autumn.—19th July to 14th November. Chiefly August.

RUFF.

Spring.—5th May to 4th June.
Autumn.—17th July to 2nd November. Chiefly first half of September.

COMMON SANDPIPER.

Spring.—23rd April to 8th June. Chiefly late May.
Autumn.—14th July to 2nd November. Chiefly September.

WOOD-SANDPIPER.

Spring.—22nd April to 26th May.
Autumn.—11th July to 17th November.

GREEN SANDPIPER.

Spring.—9th April to 21st May. Chiefly first half of May.
Autumn.—27th July to 3rd November. Chiefly late August and early September.

REDSHANK.

Spring.—10th April to 31st May.
Autumn.—20th July to 15th November. Chiefly August and September.

SPOTTED REDSHANK.

Spring.—23rd April to 29th May.
Autumn.—9th July to 27th November. Chiefly September and October.

GREENSHANK.

Spring.—10th April to 7th May and 8th June.
Autumn.—10th July to 23rd November.

BAR-TAILED GODWIT.

Spring.—16th April to 19th June.
Autumn.—14th July to 11th November. Chiefly mid-August and during September.

BLACK-TAILED GODWIT.

Spring.—10th April to 19th June.
Autumn.— 6th August to 22nd October.

CURLEW.

Spring.—3rd March to 3rd June.
Autumn.—15th August to 23rd November. Chiefly August and September.

WHIMBREL.

Spring.—15th April to 8th June. Chiefly May.
Autumn.—16th July to 29th November. Chiefly mid-August and during September.

ARCTIC TERN.

Autumn.—11th September to 29th October.

POMATORHINE SKUA.

Spring.—1st March to 11th June.
Autumn.—18th August to 27th November.

BUFFON'S SKUA.

Spring.—12th May to 14th June.
Autumn.—11th August to 29th October.

SLAVONIAN GREBE.

Spring.—5th May to 18th June.
Autumn.—9th September to 28th November.

CHAPTER VI

THE autumn migrations observed in the British Islands are all of them return-movements from summer nesting haunts to winter retreats within our areas or to warmer climes beyond their limits. They comprise local movements between various British seasonal haunts, the departure of our summer visitors, the arrival of winter visitors, and a prolonged procession of birds of passage bound southwards.

> When in a thousand swarms, the summer o'er,
> The birds of passage quit the English shore,
> By various routes the feathered myriad moves.
> —CHARLOTTE SMITH.

The various migrants usually follow the same lines of flight, but in a reverse direction, as in the spring; they depart from the same shores which witnessed their arrival a few months before; and they arrive on those from which they took their departure to proceed to their summer quarters.

The span between the last of the spring movements to northern nesting haunts and the setting in of the autumn emigrations of the British summer guests is but a short one. As early as July, while the summer is in the zenith of her glory, certain of our birds which have accomplished the duties and cares associated with the rearing of their families (or have perhaps been unfortunate in their efforts to do so), begin to leave their nesting areas and to appear in other localities more or less removed.   Some of them find their way to the coast and take leave of our shores.   In addition, there are the wanderings and migrations of young birds only a few weeks old.   It seems to me highly probable that the broods of many of our birds leave the place of their birth and go off on their own account as soon as they are fit to take care of themselves; while in the case of species which are double-brooded, the first families are often driven away by their parents.   Numbers of these roving youngsters are the offspring of migratory parents, and they form a considerable portion of the early emigrants; hence it is that a number of young birds migrate in advance of their parents.

The British migratory birds are earliest of all to leave their summer haunts.   This is to be accounted for by the fact that nesting in our areas takes place at an earlier period than in corresponding latitudes on the Continent, and much earlier, of course, than in those further north.   The late nesting season in Northern Europe explains the appearance of many birds of passage and winter visitors on our shores at dates considerably after the British summer visitors of the same species have quitted our islands.

The autumn retreat towards winter quarters is a leisurely performance in marked contrast to the feverish rush to nesting haunts in the spring.   In the early autumn there is no necessity for the migrants to hurry southwards, for food is still abundant in many places along the routes of flight.   Consequently, many migratory visitors tarry for a considerable time, some of them lingering until the falling temperature acts as a sharp reminder that it is time to seek more genial climes; then great rushes southwards and westwards set in.

Speaking generally, the insectivorous species are the earliest emigrants among the smaller birds to quit the neighbourhood of their nesting areas, and the latest to leave are the frugivorous and granivorous groups, such as the various thrushes and finches; for these the abundant crop of wild berries which carpets the northern forests and moors offer attractions which are irresistible to the majority, and induce them to remain until the season is far advanced. The latest of all the migrants to leave their summer quarters are to be found among the arctic and sub-arctic sea-birds, such as the Glaucous and Iceland Gulls, the Little Auk, etc. The date of the first arrival of the Winter Visitors and Birds of Passage depends upon the nature of the breeding season on the Continent, especially in the north—whether it be an early or late nesting season there.

At this season the numbers of the migrants are much greater than during the spring, as their ranks are swelled by numerous young recruits, most of them only a few weeks old. This circumstance, and the leisurely manner in which the fall migrations are performed, combine to make the autumn movements much in evidence and comparatively easy to observe.

To the autumn really belong those erratic appearances of the Common Crossbill on our shores, which sometimes take place ere the summer has fairly set in, as in 1909. Late in June and early in July of this year, both young and old of this species arrived in great numbers on our shores, and remained for several weeks even on the remotest of our pelagic islands, where numbers perished for want of suitable food. These remarkable movements, despite the date of their performance, must be classified among those of the autumn, for they are embarked upon after the breeding season is over, and the young birds accompany the old in these enigmatical summer wanderings.

JULY.—The comments just made on the shortness of the interval between the flow of the migratory stream towards summer quarters for nesting and the setting in of the ebb towards winter retreats, especially concern this month. The remarks on the early return-movements of the British migrants and their probable causes also belong chiefly to July.

Local Movements.—Among the most regular of the July movements are those which relate to the departure of a number of our native birds from the localities in which the summer has been spent, to seek the autumn feeding grounds, which in some cases may not be far away. Some of the earliest birds to quit their summer quarters are certain plovers and sandpipers which, accompanied by their young, return to the coast; and the fledglings of a number of "sea-fowl" leave their rocky nurseries and take to the neighbouring sea. The following species participate in these July local migrations: Grey Lag-Goose, Mallard, Shoveler, Redshank, Golden Plover, Lapwing, Ring-Plover, Oyster-catcher, Greenshank, Dunlin, Snipe, Whimbrel, Curlew, Arctic and Common Terns, Black-headed Gull, Common Gull, Kittiwake, Great Skua, Arctic Skua, Guillemot, Razorbill, Puffin, and doubtless others which have not come under my observation, for these retirements are very gradually and

PLATE IV.

The Advance of Autumn
After A. G. Högbom

The advent of autumn is proclaimed by the same degree of temperature as that which ushered in the spring. The map shows the dates of the return of isotherm 48° F., and its march southwards over the various regions of Europe.

The appearance of autumnal climatic conditions has an important bearing upon the dates on which the different kinds of summer birds quit the more northern portions of the Continent to proceed to winter retreats in warmer climes.

Plate IV

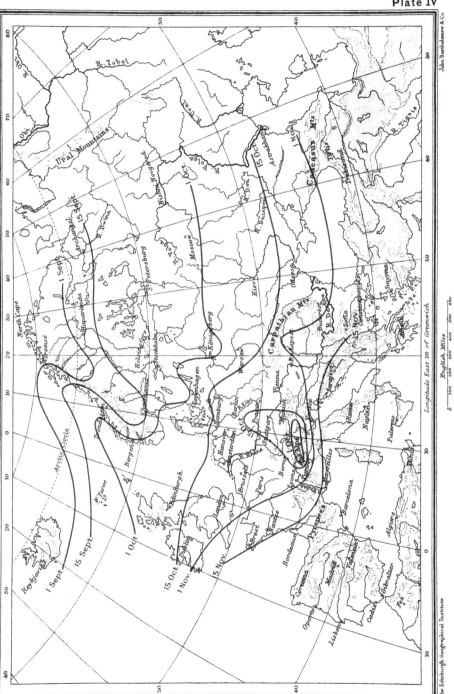

John Bartholomew & Co.

Longitude East 20 of Greenwich

English Miles

Plate 17

almost imperceptibly performed. I have noticed, however, for several years, that the invasion of their haunts
by sportsmen on the 12th of August causes certain
species to quit the moorlands, and on that day numbers
of Curlews and Golden Plovers pass over the environs of
Edinburgh, making for the shores of the Firth of Forth.

These local movements continue throughout the
early autumn, and there is no further necessity to allude
to them, since they relate to the return of the same
species which have been mentioned as seeking their
summer haunts during the spring.

Departure of Summer Visitors.—During July, especially towards the end of the month, there are records
from the light-stations which unmistakably indicate that
the departure of certain species from our islands has
already commenced. These early flittings, with a few
exceptions, must be regarded as somewhat unusual, and
possibly due to the disturbing influence of local meteorological conditions, with which, indeed, they are in many
cases correlated. During the month the following
species have been observed under circumstances which
leave little or no doubt that they were emigrating, those
marked thus * having been detected at lighthouses or
light-vessels, en route for countries beyond our shores :—
*Song - Thrush, *Blackbird, *Redbreast, Redstart,
*Wheatear, Nightingale, Willow-Warbler, Chiffchaff,
*Sedge-Warbler, *Goldcrest, *Pied Wagtail, *Meadow-
Pipit, *Swallow, Chaffinch, *Starling, *Skylark, *Swift,
*Nightjar, *Cuckoo, Corn-Crake, Common Sandpiper,
*Lapwing, *Curlew, Sandwich Tern, and Roseate Tern.
The most constant of these July emigrants are the
adult Cuckoos; and the Swift, which is one of the
latest of our summer guests to arrive, is also one of

I.   K

the first to take leave of us. It is a single-brooded bird, and does not linger anywhere in Europe, not even in its nesting haunts in countries bordering the Mediterranean, after its young are able to undertake the long journey to the southern winter homes of the species. On the 29th of July, 1887, hundreds of Swifts were observed passing southwards at Yarmouth.

Some of the birds enumerated, such as the Song-Thrush, Pied Wagtail, Meadow-Pipit, Skylark, Starling, Lapwing, and Curlew, belong to species which have been defined as partial migrants—that is to say, a proportion of the individuals composing them are migratory and essentially summer visitors, whilst the rest are sedentary, remaining in our islands all the year round.

Passage Movements.[1]—As the summer, more especially the northern summer, is yet young, it would scarcely be expected that immigrants from the north or from the Continent would arrive on our shores thus early on their return to their winter quarters. Yet a number do arrive, and the appearance of others may, perhaps, be accounted for in the following ways:—They are either (1) birds which have not proceeded so far on their spring journeys as to reach the breeding grounds, being either immature or barren birds ; or (2) if they have done so, they have either failed to obtain mates, or for some other reason have not reared families. Some of them, indeed, may not have passed much beyond British limits on their spring journeys north or east.[2] Most of the July immigrants occur late in the month,

_____

[1] For particulars of the dates of Passage Movements, see p. 129.
[2] A number of Common and Velvet Scoters, Turnstones, Purple Sandpipers, Bar-tailed Godwits, and other northern breeding species, are known to pass the summer with us as non-nesting birds.

and are seen singly or in small parties, rarely in considerable numbers or in flocks.   The majority of these earliest immigrants belong to the great Limicoline group (plovers and sandpipers), and are chiefly adult birds. Among them we find the following species :—Honey-Buzzard, Spoonbill, Grey Plover, Turnstone, Great Snipe, Knot, Little Stint, Curlew Sandpiper, Sanderling, Ruff, Dusky Redshank, Green Sandpiper, Wood-Sandpiper, Bar-tailed Godwit, and Black Tern.   There are also July records of the occurrence of the Brent-Goose, Little Gull, etc., and of a number of rare casual visitors.

A singularly interesting visitor, which makes its appearance in British waters during this month, is the Great Shearwater.   Many of these pelagic rovers, on quitting their breeding haunts in the South Atlantic Ocean, cross the equator, and move northwards to spend the southern winter months in the temperate regions of our hemisphere.

AUGUST. — *Departure of Summer Visitors.* — The chief movements in August are those of departure, and relate to the emigration of summer guests, including those belonging to species which are partially migratory. These now commence in earnest, and during the latter days of the month a number of species, and very many individuals, quit our shores.   No great movements or "rushes" are observed, for there are, as yet, no climatic incentives to cause them, but the exodus is much in evidence on the coast and its immediate vicinity.

The summer visitors of species which are partially migratory and are recorded as leaving us during this month are :—Mistle-Thrush, Song-Thrush, Stonechat, Redbreast, Goldcrest, Hedge-Accentor, Pied Wagtail,

Grey Wagtail, Meadow-Pipit, Starling, Skylark, Kestrel, Woodcock, and Curlew. Other individuals belonging to this group of migrants are noted as being on the move, though they are not necessarily passing at once beyond the British area; indeed, some of them from the more northern districts are content to remain the entire winter in the more genial portions of our isles, should the nature of the season permit them to do so.

Arrival Movements.[1]—Immigrants from the northern regions in which the summer has been passed are much in evidence on our shores during August, and are the advance guard of the hosts of Birds of Passage and Winter Visitors which are soon to follow. The former group includes northern representatives of a number of species which are summer visitors to Britain, such as the Lesser Whitethroat, Common Whitethroat, Willow-Warbler, Tree - Pipit, Pied Flycatcher, Swift, and Nightjar, all of which occur regularly but in small numbers during this month at stations north of their British breeding limit—a fact which clearly indicates that they are arrivals from countries beyond our area.

In addition to the return movements of the birds named, about thirty species, whose summer haunts lie entirely beyond the British isles (to the north or east of them), are chronicled as arriving on and traversing our shores during the month. These are: the *Greater Wheatear (Saxicola leucorrhoa)*, Barred-Warbler, Black Redstart, *White Wagtail, Blue - headed Wagtail, Hoopoe, Barnacle-Goose, Brent-Goose, Spotted Crake, *Grey Plover, *Turnstone, Ruff, Great Snipe, Jack Snipe, *Little Stint, *Curlew Sandpiper, Knot, *Purple

[1] For particulars of the dates of Passage Movements, see p. 129. For dates of arrival of Winter Visitors, see p. 157.

Sandpiper, *Sanderling, Green Sandpiper, Wood Sandpiper, Dusky Redshank, *Bar-tailed Godwit, Black-
tailed Godwit, Black Tern, Little Gull, Sabine's Gull,
Pomatorhine Skua, Great Northern Diver, and Red-
necked Grebe. Some of these also remain the winter,
among them the Barnacle and Brent Geese, Grey
Plover, Turnstone, Jack Snipe, *Knot, Purple Sandpiper, Sanderling, Bar-tailed Godwit, Redshank, Great
Northern Diver, and Red-necked Grebe. Those marked
with an asterisk occur regularly during August, while
the rest have only occasionally been recorded for the
month.

SEPTEMBER.—This is an important month for the
departure of a great variety of migratory birds from
their summer homes in far northern and north-temperate
regions. The temperature has then fallen in the
northern hemisphere, and values below freezing point
are to be found over wide areas of the arctic regions.
Departure of Summer Visitors.—In our islands,
September witnesses the height, and almost the close,
of the emigrations for winter quarters of all the numerous
species which have spent the summer with us. Our
native Swifts and adult Cuckoos are the exceptions,
for they have usually departed by the end of August.
It is true, as has already been indicated, that other birds
of nearly all the species occur later, but these are chiefly
on passage from more northern countries, where the
nesting season is much later than in our own.
    There are often considerable movements of these
emigrating British birds southwards, and finally across
the Channel, synchronous with the prevalence of
ungenial weather, which acts as a reminder to those

which no longer have ties (the tending of their young), that the time has arrived when they should bid adieu to our shores for the year.

These departures from our islands are the most difficult of all migratory movements to observe, since the birds as a rule slip away unnoticed during the hours of darkness. In order to obtain reliable information concerning them, and the various conditions under which they are performed, the author spent a month in the Eddystone lighthouse during the autumn of 1901, and his experiences there will be related in a subsequent chapter.

Passage Movements from the North.[1]—As the result of the climatic incentives to emigration already alluded to, great numbers of birds from Northern Europe arrive on our shores during the month, the majority of them en route for countries further south. One of the main features of these movements is the passage along our shores of hosts of insectivorous Passerines—Warblers, Chats, Fly-catchers, Wagtails, Pipits, Swallows, Martins, etc. The appearance of these little travellers may be said to commence in September, and practically, though not entirely, to cease with its close. When the weather conditions especially favour their passage across the northern seas, they arrive in great force, particularly during the latter half of the month, and the observer at favourably situated points has a very busy and intensely interesting time recording their kinds and noting the changed habits of many of them under the influences exerted by the exigencies of travel.

Later in the month, the members of the frugivorous

[1] For particulars of the dates of Passage Movements, see p. 129. For dates of arrival of Winter Visitors, see p. 157.

and granivorous sections of the order Passeres—Song-Thrushes, Blackbirds, Ring-Ouzels, Redwings, Fieldfares (a few), Chaffinches, Bramblings, Siskins, Mealy Red-polls, and Skylarks—*begin* to appear on our shores, usually in small numbers, but sometimes in flocks containing many individuals. All the species named, except the Ring-Ouzel, comprise winter guests with us, as well as visitors bound for more southern retreats.

The arrivals of northern migrants belonging to the great group of birds popularly known as "waders," and including the plovers, sandpipers, and snipes, continue throughout the month, and the movements of many of the species—the Grey Plover, Little Stint, Curlew Sandpiper, Sanderling, Ruff, Bar-tailed Godwit, and Whimbrel—reach their maximum importance. On the other hand, some, such as the Woodcock, only occasionally appear among us as stragglers, or as scouts in advance of the vanguard. Many species of ducks and several of geese also make their appearance from the north; among them the White-fronted Goose, Bean-Goose, Pink-footed Goose, Barnacle-Goose, Brent-Goose, Mallard, Teal, Wigeon, Scaup, Golden-eye, Long-tailed Duck, and Common and Velvet Scoters.

Among the more interesting of the September passage-migrants are the Greater Wheatear, Red-spotted Bluethroat, Barred-Warbler, Yellow-browed Warbler, Icterine Warbler, Pied Flycatcher, Grey-headed Wag-tail, Lapland Bunting, Ortolan Bunting, Honey-Buzzard, Osprey, Wood-Sandpiper, Dotterel, Great Snipe, and Dusky Redshank.

Much information regarding the passage movements of these migrants from the north will be found in the studies devoted to Fair Isle, St Kilda, and the Flannan Islands, the former of which occupies a singularly favourable geographical position for their observation.

The steady flow of migrants southwards, towards and along our shores, is accelerated during the latter half of the month, when a series of "rushes," the result of weather influences in the north, usually take place.

Arrivals from the East.—The autumn immigrations hitherto considered are those of birds which come to us from Northern Europe, Iceland, and Greenland; and from North-western Siberia in the case of the Curlew Sandpiper, Yellow-browed Warbler, and others. There now remain for consideration those migrants (birds of passage and winter visitants) which reach us by an east-to-west flight across the southern waters of the North Sea, and arrive on the south-eastern coast of England between the Wash and the Channel.[1]

These movements commence about the middle of September, when Skylarks, Starlings, Tree-Sparrows, Chaffinches, etc., are observed at the light-vessels off the coast streaming day after day towards the English shores, chiefly during the daytime. Later, Grey Crows, Rooks, Jackdaws, and Lapwings figure largely among the throng moving westwards.

This important and singularly interesting stream of migration received my personal attention in September and October 1903, when I spent a month on board the Kentish Knock lightship, for its special investigation. The results of these observations will be given in Chapter XVIII., Vol. II., which renders it unnecessary to treat of them further in this place.

[1] For the geographical aspects of this migration route, see p. 83, and map (Plate II.).

OCTOBER.—During October, the autumn migrations reach their greatest magnitude, but chiefly consist of arrival and passage movements from the north and east.

Passage Movements and Arrival of Winter Visitors.[1] —The rapid lowering of the temperature in Northern Europe, which is one of the climatic features of October, causes vast numbers of summer residents in those wide-extending regions to move southwards, with the result that our own and other countries are flooded by successive rushes of migrants, the magnitude of which is unequalled during any other month of the year. The arrivals on the south-east coast of England, from the opposite shores of the Continent, also reach their maximum volume in October, and add considerably to the extent and variety of the movements on that exceedingly busy section of our coast-line.

Day after day during the month, and especially in its latter half, when the weather conditions are favourable, winter visitors pour into the British Islands from both the north and east; while more, very many more, migrants, mostly of the same species, rush along our coasts as birds of passage en route for more southern winter quarters.

The typical October migrants from the north are mainly composed of fruit- and seed-eaters among the Passeres—the various species of thrushes and finches—of Hedge-Accentors, Grey Crows, Starlings, Skylarks, Short-eared Owls, Swans, various species of Geese, Ducks, Plovers (including the Woodcock) and Sandpipers, Snipe, Divers, and Grebes. Less abundant, but

---

[1] For particulars of the dates of Passage Movements, see p. 129. For dates of arrival of Winter Visitors, see p. 157.

not less interesting, are the Mealy Redpoll, Great Grey Shrike, Shore-Lark, and Rough-legged Buzzard.

From the east prodigious numbers of Rooks, Jackdaws, Starlings, Chaffinches, Greenfinches, Tree-Sparrows, and some Mistle-Thrushes rush across the North Sea from the Dutch coast to the shores of East Anglia, the mouth of the Thames, and the eastern littoral of Kent.

During the early days of the month, the rearguard units of the great army of the insect-eating birds— the warblers, flycatchers, wagtails, pipits, swallows, etc. —and also of the waders—Little Stints, Curlew-Sandpipers, Common Sandpipers, Green Sandpipers, and Ruffs—pass southwards. While among the rarer visitors are the Little Bunting, Siberian Chiffchaff, Yellow-browed Warbler, Red-spotted Bluethroat, Black Redstart, Red-breasted Flycatcher, Dusky Redshank, and Wood Sandpiper.

It would be wearisome to enter into further particulars regarding various species of regular migrants (no less than 140 in number) on the move during the month; but those who wish to have further details will find much information on referring to pages 129-140, and to the chapters devoted to Fair Isle, the Flannans, and the Kentish Knock.

Emigration of Summer Visitors.—The departure of those birds which have spent the summer with us has already practically ceased, though some of them which rear more broods than one during the season, such as the House-Martin and others, are often detained by late families sometimes well into the month.

Many birds of species which are summer visitors to our isles do indeed occur, chiefly on the coasts and in their vicinity, during the month. These are not usually emigrating British birds, but travellers from other countries (chiefly northern ones), visiting our shores as birds of passage on their way to distant winter retreats, and as such their movements have already been noticed.

NOVEMBER.—The normal conditions of the northern winter now prevail on the Continent.

The migratory movements of the month are a continuation of the arrival from the north and east of winter visitors to our isles, and of birds of passage on their way further south. These birds are of the same species as those described as being typical October migrants, and they appear in considerable numbers down to the third week of the month, after which stragglers only are usually observed.

The Whooper and Bewick's Swans, various species of Duck, Slavonian and Red-necked Grebes, and the Little Auk, appear in force; and it is the month for the occurrence of the Waxwing, when that beautiful "Bohemian" appears among us.

Certain other species have been known to occur in November, but their appearance so very late in the season must be regarded, in most cases, as somewhat exceptional; these are the Wheatear, Redstart, Black-cap, Willow-Warbler, Chiffchaff, White Wagtail, Tree-Pipit, Swallow, House-Martin, Sand-Martin, Hoopoe, and Common Sandpiper.

Autumn Casual Visitors.—The autumn occurrences of casual visitors are far in excess, both in numbers and kinds, of those recorded for spring. This is readily accounted for by the number of young birds which are now undertaking their first migratory journeys, and are prone to stray from the accustomed routes to winter quarters followed by their kind. These erring youngsters form the great majority of the waifs visiting us at this season. Casuals of no less than 120 species have from time to time occurred in the British Islands.

During the great autumn movements, the emigratory ones in particular, migrants are frequently observed simultaneously on all our shores; and under certain peculiar weather conditions, to be explained in the chapter devoted to migration-meteorology, there are immigratory and emigratory movements simultaneously in progress.

The autumn migrants (including the birds of passage) as a rule arrive on and leave our shores during the hours of darkness (excepting those which come direct from the east); but some emigrant Swallows, Wagtails, Starlings, etc., cross the Channel during the morning, commencing soon after daybreak, and ceasing to do so about mid-day.

It is chiefly during the great movements of the late autumn that we hear migratory birds passing overhead on dark nights, when proceeding to their inland winter quarters. The object of their calls is probably to enable them to keep in touch with each other. On these occasions they are evidently much alarmed or excited by the lights of our cities and towns, and hence a babel of tongues which disturbs the quietness of the night, recalling to mind Longfellow's beautiful lines—

> I hear the cry
> Of their voices high,
> Falling dreamily through the sky,
> But their forms I cannot see.

APPENDIX.—DATES OF THE ARRIVAL OF WINTER VISITORS

In preparing the following data, preference has been given to observations made at stations where the arrival of the immigrants was most likely to be at once detected, and where also they are not likely to be confounded with the emigratory or local movements of our native birds of the same species, which may be in progress at an identical date.

The arrival on our shores of these Winter Visitors and of the Birds of Passage (en route for their cold-weather retreats in countries south of the British Isles) often takes place simultaneously ; and when the arriving birds belong to the same species, it is impossible to determine to which of these two great categories of British migrants they belong.

Species.	Early Records.	Usual Date of Arrival.
Rook	1st September .	29th September to 29th November.
Grey Crow	5th August .	7th October to 9th November.
Jackdaw	6th October .	17th October—
Starling.	20th September .	24th September to 22nd November.
Chaffinch	3rd September .	23rd September to 11th November.
Brambling	5th September .	24th September to 19th November.

DATES OF THE ARRIVAL OF WINTER VISITORS—*continued*

Species.	Early Records.	Usual Date of Arrival.
SISKIN	28th August	24th September to 26th November.
MEALY REDPOLL	16th August	13th October to 28th November.
TREE-SPARROW	23rd September	3rd October to 9th November.
GREENFINCH		13th October to 26th November.
YELLOW BUNTING		8th October to 31st October.
REED-BUNTING		20th September to 10th October.
LAPLAND BUNTING	25th August	7th September—October.
SNOW-BUNTING	7th September	12th September to 10th November.
SKYLARK		17th September to 15th November.
SHORE-LARK	20th September	15th October to 17th November.
MEADOW-PIPIT		11th September to 8th October.
GOLDCREST (CONTINENTAL RACE)	8th September	23rd September to 16th November.
GREAT GREY SHRIKE	1st September	10th October to 20th November.
WAXWING	9th October	November, December, and January (irregular).
MISTLE-THRUSH	21st September	7th October to 20th November.

SONG-THRUSH (CONTINENTAL RACE) .	4th September .	8th September to 25th November.
REDWING .	15th September .	23rd September to 16th November.
FIELDFARE .	2nd September .	7th October to 22nd November.
BLACKBIRD .	14th September .	20th September to 20th November.
REDBREAST (CONTINENTAL RACE) .	8th September .	20th September to 19th November.
BLACK REDSTART .	27th August .	20th September to 6th November.
HEDGE-ACCENTOR (CONTINENTAL RACE) .	20th August .	19th September to 9th November.
WREN .	8th September .	15th September to 12th October.
GREAT-SPOTTED WOODPECKER (CONTINENTAL RACE) .	8th September .	21st September to 27th November.
SHORT-EARED OWL .	2nd August .	28th September to 8th November.
LONG-EARED OWL .	5th August .	6th October to 13th November.
ROUGH-LEGGED BUZZARD .	15th August .	6th October to 12th November.
KESTREL .	9th August .	2nd September to 1st November.
HERON .	8th July .	2nd September to 29th September.
GREY LAG GOOSE .	21st August .	24th September to 13th November.
WHITE-FRONTED GOOSE .	.	29th September to 28th October.
BEAN GOOSE .	11th September .	5th October to 3rd November.
PINK-FOOTED GOOSE .	24th August .	18th September to 18th October.
BARNACLE-GOOSE .	10th July .	16th September to 24th November.

DATES OF THE ARRIVAL OF WINTER VISITORS— *continued*

Species.	Early Records.	Usual Date of Arrival.
BRENT-GOOSE	12th August .	20th September to 21st November.
WHOOPER SWAN	6th October .	21st October to 14th November.
BEWICK'S SWAN	10th October .	3rd November to 30th November.
MALLARD	. .	12th September to 5th November.
GADWALL	26th August .	23rd September to 29th October.
SHOVELER	5th September .	....
PINTAIL	11th August .	17th September to 11th October.
TEAL	. .	6th September to 28th November.
WIGEON	5th August .	9th September to 16th November.
POCHARD	. .	3rd September to 14th November.
TUFTED DUCK	. .	15th September to 13th October.
SCAUP DUCK	1st August .	13th September to 9th November.
GOLDENEYE	26th August .	16th September to 16th November.
LONG-TAILED DUCK . . .	24th July .	26th September to 31st October.
COMMON SCOTER . . .	5th August .	10th September to 9th October.
VELVET-SCOTER	. .	16th September to 18th October.
GOOSANDER	21st August.	25th September to 31st October.

MERGANSER	.	15th September to 20th October.
SMEW	6th August	12th October to January and February.
RING-DOVE	.	24th September to 15th November.
WATER-RAIL	.	24th September to 14th November.
SPOTTED CRAKE	15th August	20th September—
GOLDEN PLOVER	5th August	12th August to 12th November.
GREY PLOVER	1st July	10th August to October.
LAPWING	.	7th September to 12th November.
RINGED PLOVER	.	7th September to 14th November.
TURNSTONE	20th July	12th August to 15th November.
OYSTER-CATCHER	.	20th September to 12th October.
GREY PHALAROPE	1st August	18th September to 4th November.
WOODCOCK	23rd September	9th October to 19th November.
JACK SNIPE	20th August	15th September to 24th November.
COMMON SNIPE	17th August	6th September to 26th November.
DUNLIN	22nd July	9th August to 14th November.
PURPLE SANDPIPER	4th August	12th August to 17th October.
KNOT	16th July	12th August to 18th November.
SANDERLING	14th July	7th August to 27th September.
REDSHANK	13th July	7th August to 2nd November.

I.

L

Dates of the Arrival of Winter Visitors—*continued*

Species.	Early Records.	Usual Date of Arrival.
Bar-tailed Godwit	14th July	10th August to 11th November.
Curlew	.	Mid-August to 12th November.
Glaucous Gull	28th September	5th October.
Iceland Gull	11th July	9th October.
Great Northern Diver	12th July	16th September.
Black-throated Diver	.	25th October to 26th November.
Red-throated Diver	13th August	12th September.
Red-necked Grebe	17th August	20th September.
Slavonian Grebe	27th August	9th September, October.
Eared Grebe	25th July	September—November.
Little Auk	15th July	17th September to 7th November.

NOTE.—It is most difficult to determine the dates between which a number of marine species arrive off our shores to pass the winter. Their appearance seldom comes under immediate notice.

CHAPTER VII

ROUND THE YEAR IN THE BRITISH ISLES : WINTER

THE migrations observed in the British Isles during the winter consist of: (1) movements between various parts of our area; (2) emigrations from our shores; and (3) some immigration from the Continent. All these are due to the effect of meteorological conditions of an adverse nature upon the food supplies, either within the British area, or on the Continent.

Not very long after the last of the summer guests have bidden us *au revoir* until another spring, and while the autumn visitors have scarcely ceased to arrive to spend the winter, another set of bird-movements may, sooner or later, be ushered in—forced migrations to escape the rigors imposed by the season.

The date on which such disturbances and their consequent evictions take place, depends upon the climatic aspect of the season. They may commence in November, and are not unknown to March; but December and January are the main months for their occurrence.

Few winters are entirely free from cold spells, and some are notable for their severity. When the land becomes icebound, or a pall of snow covers its surface, it is impossible for many birds to obtain food, especially for those which seek it on the ground or in the marshes. A bountiful supply of food is more than ever a necessity when great cold prevails, and an exodus from the area affected becomes imperative when it fails. During these severe periods the inland waters are sealed with ice, and many kinds of Duck, Water-Rails, Coot, and other aquatic species move to the coast and its estuaries.

The remarkable distribution of temperature in our islands during the winter has an effect on the retreats sought, within their limits, when these stressful periods set in. The diminution of our winter temperature is from west to east rather than from south to north—a fact which is considered to be one of the most singular among the climatic phenomena in the world. If a map showing the December isotherms be consulted, it will be found that the Shetlands, the Hebrides, Skye, Cantyre, Galloway, the western angle of England from Holyhead to Brighton, have a mean temperature of above 40° F. Milder still are the south-western angles of England, Wales, and Ireland. Experience seems to have taught the birds that these western and south-western portions of Great Britain and Ireland are likely to offer the most genial retreats when the distressful outbursts of cold and snow set in, for they are the main areas sought by the feathered refugees.

Should the winter be uniformly mild, as that of 1881-82, the birds resident or sojourning with us remain practically stationary—that is to say, their wanderings in search of food do not assume the form of migrations.

British Local and Emigratory Movements.—As soon as frost sets in, particularly if it be accompanied by snow, Mistle - Thrushes, Song - Thrushes, Redwings, Fieldfares, Blackbirds, Greenfinches, Linnets, Starlings,

Skylarks, Meadow - Pipits, Ring - Doves, Lapwings, Golden Plover, Woodcock, Snipe, Dunlin, Curlew, etc. (and in some seasons even Rooks, Magpies, and Snow Buntings), remove themselves from its baneful influence, numbers of them flitting during the hours of darkness, when, as Longfellow has observed, we

> Hear the beat
> Of their pinions fleet,
> As from the land of snow and sleet
> They seek a southern lea.

Sometimes suitable retreats may be found near at hand, and then the movement is but of a local nature. When this is the case, and the period of stress is short, the evicted birds, or some of them, soon return to their usual haunts. Should, however, these adverse conditions prove severe and the area affected widespread, the movements assume like proportions, and the refugees are many. On such occasions the milder conditions usually afforded by the coast are sought, especially our western sea-board and its off-lying islands, including even the far-off Flannans, and some of the emigrants seek asylums in Ireland. If the cold be exceptionally severe and general over the British area, and its prevalence seriously prolonged, and should it be accompanied by heavy falls of snow, then a great exodus follows, and the numbers of the fugitives become vast and surprising. The movements are again chiefly towards the coast: along the east and west shores the fleeing birds rush southwards to reach south-western England and Ireland, where more genial conditions usually, but not always, prevail; while many cross the English Channel for Southern Europe. Many also pursue an overland course in a southerly direction, and

I.   L 2

congregate on the south coast of England. Numbers
are known to quit Ireland when the Sister Isle is included
in the sphere of an arctic winter.   One of the features of
such great retreats before the storm—before heavy
snow in particular—is the rush westwards, along the
south coast of England and its vicinity, of hordes of
Skylarks and Starlings, accompanied by numbers of
Song - Thrushes, Redwings, Fieldfares, Blackbirds,
Linnets, Lapwings, and doubtless other species.   Sometimes these rushes last for several days, and the number
of Skylarks observed is simply astounding, and probably includes immigrants of this, and of the other
species named, from Western Central Europe, which
have arrived on the south-east coast of England after
an east-to-west passage across the North Sea.

The birds evicted from Great Britain enter Ireland
at various points.   Those leaving Scotland arrive on
the north-east coast; those from Wales on the east
coast; while those arriving from south-western England
(some of them after traversing the south coast) enter the
country at its south-east angle, after a passage across
St George's Channel—the route chiefly used by the
spring, and many of the autumn, visitors to the Sister
Isle.

In Ireland there are many local winter movements
due to the pressure of weather, and then the west coast
is much visited, the south-western counties of Cork and
Kerry especially being resorted to.   Winter emigration
must, however, be regarded as an exceptional phenomenon
in Ireland, for some section or another of its area usually
affords an asylum in all save the severest of seasons.

During the severe weather of January 1881, when
as many as 25° of frost were registered in the west of

Ireland, Mr Warren noted that the snow and low temperature drove Rooks, Magpies, Blackbirds, Thrushes, Fieldfares, Redwings, Skylarks, Meadow-Pipits, Starlings, Hedge-Accentors, Redbreasts, Stonechats, Chaffinches, and Yellow Buntings to the shore to search for food amongst the stones and seaweeds.

In the terrible December of 1882, and the winters of 1890-91 and 1894-95, vast numbers of refugees perished in such usually sure retreats as the Scilly Isles and Valentia, among others Snow-Buntings, the hardiest of all the small birds. Regarding the effects of the great frost of 1895, the lighthouse keeper at Samphire Island, off the west coast of Kerry, observes[1] that on 7th February vast numbers of Starlings, Skylarks, Thrushes, and Redwings were going south all day in heavy snow, and that "the island was literally covered with them as they rested upon it, and at times they would darken the sky." Next day they were still going south in "one continuous flight; they were all very exhausted, and numbers of Starlings fell. I never saw such a constant rush of birds." On the 9th, great numbers were again noted resting; on the 11th and 12th, numbers dead; on the 13th, he tells us: "I never saw such a rush of birds as there has been for the past week . . . they were composed of Lapwings, Golden Plover, Starlings, Redwings, Thrushes, Skylarks, Rooks, and Fieldfares, and some Grey Linnets. . . . It is pitiful to see the hundreds of birds dead and dying all about the island, particularly Starlings; the sea is covered with them and Gulls feeding on them, although vast numbers of Gulls have died."

[1] *Report on the Migration of Birds on the Irish Coast in* 1895, p. 477.

Later in the season, the weather-evicted birds are those which have ventured an early return to summer quarters, especially such as have sought the higher and more exposed situations. The birds chiefly affected are ground-loving species, such as Lapwings, many of which perished in Scotland during the storm of 26th to 30th March 1901 ; and snow is the main cause of their discontent. As a rule, the unfavourable conditions do not then prevail for long, and the fugitives soon seek again their nesting grounds ; some of the birds, however, are loth to quit their chosen haunts, and remain to brave the storm, particularly if it be a late one, and then many perish. With few exceptions these compulsory retirements cease during the first half of March.

The fact that these winter migrants largely seek the coast and its neighbourhood, when evicted from their British inland haunts, has often caused them to be regarded as arrivals from the Continent. Some immigrants from abroad do appear on particular sections of our coasts, and to these allusion will presently be made.

The cold-weather movements are observed in progress both at night and during the daytime ; but the emigrants crossing the English Channel for South-western Europe make the passage chiefly during the hours of darkness.

Immigration from the Continent, etc.—During very severe winters on the Continent and in the northern seas, a number of refugees visit us, some of them to remain, others on their way, perhaps, to more southern retreats.

Not many land birds are derived from the north in such seasons, for those that are not winter residents leave in the autumn. A few Fieldfares, Thrushes, Woodcock, Snipe, and Plovers have during exceptionally severe winters made their appearance in Fair Isle and

Orkney under circumstances which lead one to suppose that they were arrivals from more northern localities, perhaps from the Shetland Isles, where in ordinary seasons they would have spent the winter.

Land birds certainly do appear on the south-eastern coast of England, after a westerly passage across the North Sea, during periods of exceptional severity in Western Central Europe, the birds which then arrive including Starlings, Skylarks, Song-Thrushes, Lapwings, and Bitterns. Some of these immigrants may remain in eastern England, while others certainly pass westwards to seek the more genial retreats usually to be found there. Crowds of wild-fowl (various ducks and geese) also migrate from the Dutch coasts when their shallow waters are frozen, and seek the opposite shores of England.

The winter of 1894-95 was memorable for its arctic severity in the British Isles, and in the north of Europe the cold was phenomenal. This resulted in our seas being thronged with such an assemblage of northern sea-fowl as had not been witnessed within the memory of living naturalists. This cold period (which has already been alluded to in connection with British local movements) set in on 30th December, and prevailed, with a week's respite, until 5th March. The weather was very severe from 26th January to 19th February, when even in the south-west of England and Ireland the temperature was 10° below normal values, and in the east of Scotland 17° below them. A series of severe gales from the north drove vast numbers of Little Auks before them, and wrecked them in thousands from the Shetlands southwards along the entire east coast of Britain, and many were blown quite across Scotland and far inland in England. Other northern species were also driven south by the extraordinary severity of the season ; among them Brünnich's Guillemot, a resident in the arctic seas which had once previously—and then a single example only—been recorded as having visited our shores. Several specimens of this interesting visitor were obtained as far south as the Yorkshire coast, and one was captured inland in Cambridgeshire. Waxwings, Smews, Little and Glaucous Gulls, Red-necked Grebes, and Great Northern Divers were also exceptionally abundant.

Among the rarer species which make, or have made, their appearance chiefly in winter, are White's Thrush, Tengmalm's Owl, the Snowy Owl, Hawk-Owl, Greenland Falcon, Iceland Falcon, Red-breasted Goose, Snow-Goose, Buffel-headed Duck, Harlequin Duck, Hooded Merganser, and Ivory Gull.

CHAPTER VIII

WEATHER INFLUENCES : THE METEOROLOGY OF
BIRD-MIGRATION [1]

It has long been realised that birds are extremely sensitive to atmospheric conditions, and that they discern approaching meteorological changes. It is on this account that in popular weather-lore, the sayings and proverbs based upon the actions of these feathered barometers are so exceedingly numerous; and probably many of them date from Roman times.

Migratory birds are specially influenced, since their movements are undoubtedly correlated at all seasons with the weather, which has a controlling effect over their performance, either as an incentive to embark upon their aerial voyages, or as the greatest of barriers to their performance.

A knowledge of the meteorological conditions associated with bird-movements is extremely important, and I have often found it to contribute in a remarkable way to an interpretation of their precise nature.

When studying the meteorological aspects of bird-

[1] It has been my good fortune to have the meteorological statements made in this chapter verified by W. N. Shaw, Esq., LL.D., D.Sc., F.R.S., etc., the Director of the Meteorological Office, in accordance with whose criticism and advice it has been amended. I desire to acknowledge fully Dr Shaw's most valued assistance.

migration, it is essential that the weather conditions prevailing in the area in which the movement had its origin should be taken into consideration. Here alone, as a rule, has the weather any bearing upon the movement under investigation; the weather prevailing in places reached after a more or less extended flight is nothing to the purpose. Thus we must look to Continental conditions for an explanation of the arrival of immigrants on our shores, in both spring and autumn; and to our home weather for the key to the cause of the departure of emigrants from the British area at all seasons.

As the result of much careful study of a vast amount of material, migrational and meteorological, some very important facts have been made abundantly manifest, foremost among which are the following :—

(1) That the type of weather which prevails when the winds range from south to east, and especially when they are south-east, is the most *favourable* for migratory movements between the British Islands and Continental Europe, in both spring and autumn. (2) That the weather conditions prevailing when the winds vary from west to north are the most *unfavourable*, both for immigration to and emigration from the British Isles.

Year after year, the accuracy of these findings has been repeatedly impressed upon me during my practical work at widely separated observing-stations both at home and abroad; and they have also become known to a number of observers (lighthouse-keepers and others) who have assisted me in the investigations.

With these well-ascertained facts as a basis to work upon, let us proceed to inquire what the types of weather are which have such remarkable influences over the

PLATE V.

A large Continental anticyclone lies to the east of the British Isles, and
extends to our shores. To the west the pressure decreases, especially off
the Atlantic Coast of Ireland and over Iceland. Wind, a south-easterly
current over the United Kingdom.

The south-easterly type of weather is the most favourable of all for intermigration between Great Britain and the Continent of Europe in both spring
and autumn. During its prevalence the great migratory movements, termed
"rushes," are performed.

Fine weather is then in the ascendant over the countries whence the
great majority of our bird-visitors come to us in the autumn, and to which
they return on leaving us in the spring. The North Sea is spanned, at both
seasons, by weather which is most promising for the voyagers.

On the other hand, the western low-pressure area, with its unsettled
conditions, extends to Iceland, and is against intermigration between the
British Isles and that great north-western island. (See also p. 173.)

PLATE V.

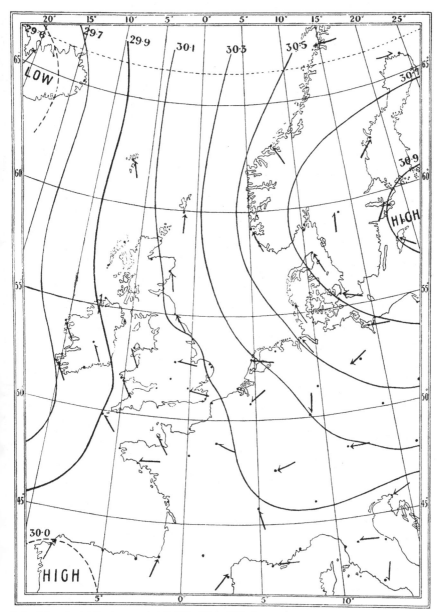

Reduced from Special Chart prepared at the Meteorological Office.]

SOUTH-EASTERLY TYPE OF WEATHER.

Isobars are shown by black lines, with indications of the height of the barometer.

WIND.—Arrows flying with the wind, show direction and force, thus :—

⟶ = forces 5 to 7 (velocity, 28 to 40 miles per hour).

⟶ = forces 1 to 4 (velocity, 8 to 23 miles per hour).

migrants or would-be migrants. It must first be re-marked, however, that the *direction* of the wind has in itself nothing to do with the results described. The winds and the performance, or non-performance, of the migratory movements are the effects of a common cause—namely, the particular type of weather prevailing at the time, which may be favourable or unfavourable for the flight of birds to or from our islands.

Favourable Weather Conditions. — The favourable conditions prevail when a large anticyclone has its centre in North - Western Europe, covers the North Sea, and extends to the British Isles. This would give light gradients (isobars far apart), gentle winds, and, generally speaking, fine weather between the British Isles and Scandinavia. Such a weather period would be eminently favourable for the passage of birds along one of the most important lines of flight for migrants between the Continent and our islands. On these occasions, too, the barometric pressure being highest to the east of our islands, the wind, according to general rule, would be between south and east over the western shores of the North Sea (see Plate V.).

When such weather conditions prevail in the autumn, much emigration takes place from North-Western Europe towards the British coasts, where the migrants make their appearance in great numbers. When the same type of weather obtains in the spring, great emigrations from our shores set in towards the north and east, of de-parting winter guests, and birds of passage, returning to their summer quarters. The pulse, if one may be allowed the term, of the movements is quickened from time to time through the stimulating influence of temperature—a fall in autumn, a rise in spring.

The greatest movements, or "rushes," as they have not inappropriately been designated, in both spring and autumn, are those which follow the advent of favourable weather on the passing away of a more or less prolonged spell of adverse conditions. These unfavourable periods in the autumn are not unfrequently characterised by great ungeniality, and this, no doubt, gives the summer visitors warning that the time has arrived for seeking milder climes in which to pass the winter. Upon the duration and severity of the unfavourable conditions for migration depends the magnitude of the exodus which follows when the anticyclone removes the barrier, releases the flood of pent-up migrants, and furnishes ideal conditions for flight over the North Sea. When such a sequence of weather changes takes place, it is not surprising that vast rushes southwards follow, and that our shores receive great waves of migrants, sometimes for several successive nights.

The gentle barometric gradients for east to south winds, with their fine weather, do not always, however, entirely bridge, as it were, the North Sea between the Continent and Britain. In the autumn, the favourable conditions which induced the birds to quit North-Western Europe may not extend to the British shores (nor to those of 'Scandinavia in the spring). Indeed, it not unfrequently happens that the bird-voyagers pass into more or less adverse conditions—high winds or heavy rain, or both—ere our shores are reached, and arrive in a correspondingly exhausted state. Occasionally many perish during the latter stages of the passage, and their bodies are cast upon our eastern coasts in considerable numbers. In connection with these unfortunate autumn flights across the North Sea, an examina-

tion of the weather charts affords a simple explanation
—namely, that though the weather in Scandinavia was
entirely favourable for emigration, the conditions prevailing on the British coasts were adverse, owing to the too
close proximity of a western low-pressure centre. On
the location and character of this cyclonic centre depends
the nature of the weather in the immediate neighbourhood of our shores. If it be too close to Britain, or if the
depression be exceptionally deep, then strong winds are
experienced on our eastern shores, with the unfortunate
results to the bird-travellers just described. It must be
remembered, too, that the winds are not necessarily light
except in the central region of an anticyclone. On the
rims we sometimes get strong winds. If the south-
western rims were off the east coast of Britain similar
disasters would befall the approaching migrants.

It has already been mentioned (p. 156) that under
certain weather conditions in the autumn, both arrival
and departure movements are in progress simultaneously
on our shores. On these occasions, it has been ascertained that the anticyclone in North-Western Europe
covers a wide area and extends beyond the limits of the
British Isles, and hence weather is favourable for
departure movements from our islands. When such is
the case, there occurs, simultaneously, an inpouring of
birds on to our shores from Northern Europe, and an
outpouring from Britain of emigrants of many species.

The duration of favourable spells for bird-migration,
in any area, is sooner or later broken by the setting in
of a cyclonic period, with more or less unfavourable
conditions, which may curb, or make impossible, the
progress of the seasonal movements.

The following experience (one of many) at Fair Isle

well illustrates the influence of such favourable conditions.   On 11th May 1910, after a prolonged period of
northerly and westerly winds, during which few or no
migratory movements were observed, the wind changed
to south and gradually to east, and varied between these
points until the 22nd.   The effect of this change was
marvellous ; the long-held-up Birds of Passage, on their
way northwards, arrived during the night of the 11th
or the early hours of the 12th in extraordinary numbers,
and afforded the Duchess of Bedford and myself a
most interesting as well as a busy time.   During this
period numbers of species came under our notice, but
doubtless many others escaped our most assiduous
attentions by resorting to the face of the great cliffs,
where it was impossible to detect them ; those observed
were Mistle-Thrushes, Song-Thrushes, Fieldfares, Ring-
Ouzels, Wheatears, Greater Wheatears, Whinchats,
Redstarts, Black Redstarts (two), Red-spotted Bluethroats (a number), White-spotted Bluethroat (male),
Redbreasts, Whitethroats, Lesser Whitethroats, Black-
caps, Chiffchaffs, Willow - Warblers, Wood - Warblers,
Sedge - Warblers, Hedge - Accentors, White Wagtails,
Blue-headed Wagtails, Grey-headed Wagtails, Tree-
Pipits, Meadow - Pipits, Red - backed Shrikes, Pied
Flycatchers, Spotted Flycatchers, Swallows, House-
Martins, Sand - Martins, Siskins, Chaffinches, Bram-
blings, Ortolan Buntings, Reed-Buntings, Goatsucker,
Wrynecks, Hoopoe, Cuckoos, Long-eared Owl, Merlins,
Kestrels, Ring Doves, Corn-Crakes, Golden Plovers,
Lapwings, Great Snipe, Common Snipe, Dunlins,
Common Sandpipers, Green Sandpiper, Redshanks, and
Whimbrel.

The movements from east to west, and *vice versâ*,

PLATE VI.

CHART OF NORTH-WESTERN EUROPE SHOWING TYPICAL WEATHER
CONDITIONS, WITH NORTH-WESTERLY WINDS, OVER THE BRITISH
ISLES

A depression lies to the eastward of and extends to our Isles; while
pressure is highest to the north-west and south-west. Wind, circulating
cyclonically round the pressure minimum, being north-westerly and
northerly over the British area and the North Sea region generally.

This type of weather is the most unfavourable of all for bird-movements
between the British Isles and the Continent. During its prevalence the area
from which we derive the great majority of the migrants in the autumn,
and often our own shores from which they return in spring, are under
unsettled weather-conditions, and hence adverse for these movements.

On the other hand, the conditions prevailing over Iceland and the
countries to the south-west of us are not unfavourable for migration from
those areas to the British Isles. (See also p. 177.)

PLATE VI.

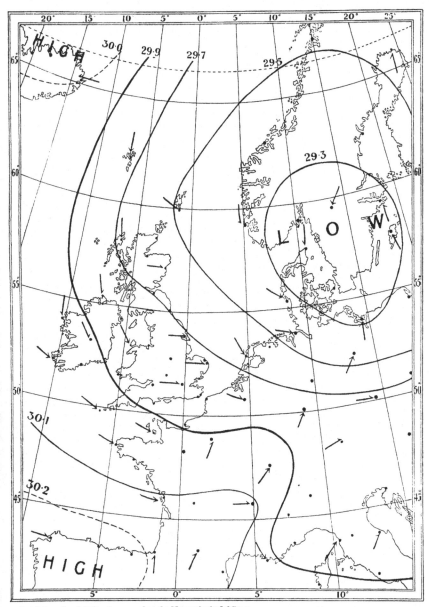

Reduced from Special Chart prepared at the Meteorological Office.]

NORTH-WESTERLY TYPE OF WEATHER.

Isobars are shown by black lines, with indications of the height of the barometer.

WIND.—Arrows flying with the wind, show direction and force, thus :—

⟶ = forces 5 to 7 (velocity, 28 to 40 miles per hour).

⟶ = forces 1 to 4 (velocity, 8 to 23 miles per hour).

across the southern waters of the North Sea are undertaken under the favourable meteorological conditions which have just been explained—namely, the prevalence of fine weather at the area of embarkation, and extending to the east in spring and to the west in autumn (see Plate V.). Some observations on the autumn weather-influences favourable and unfavourable to passages along this interesting line of flight will be found in Chapter XVIII., which is devoted to an account of my bird-watchings on board the Kentish Knock Lightship.

Unfavourable Weather Conditions.—When we come to inquire into the weather which prevails when the winds range from west to north, which is the most unfavourable type of all for the passage of migratory birds to and from the British Isles, the reason becomes at once apparent. It results from the presence of a cyclonic or low-pressure area to the north-east or east of the British Isles—a type of weather which is fatal to intermigration between North-Western Europe and Britain, because unfavourable meteorological conditions then prevail over the area whence we derive the majority of our migrants in the autumn, and over our isles. It is also against the return movements from the British shores in spring. On the other hand, this type of weather may not be unfavourable for migration from Iceland to Northern Britain in autumn; nor for arrivals in England from countries to the south-west of us in spring (see Plate VI.).

During the prevalence of these westerly winds, if the weather be not severely unsettled, some migration from the Continent takes place, but no general movements

I.   M

are recorded — the birds arrive in small numbers and
come little under notice.

There are other types of weather which, in like
manner, favour particular migratory movements, but are
unfavourable for others. Thus the North-Easterly Type
is the most favourable for migration between Iceland
and the British Isles, but is not suited for movements
between our shores and those of countries lying to the
south and south-east of them (see Plate VII.). The
South-Westerly Type is conducive to intermigration
between England and Central and Southern Europe,
but presents barriers to passages between our Isles and
Northern Europe and Iceland (see Plate VIII.).

Winds.—The importance of winds in connection with
bird-migration has been much over-estimated, and their
bearing upon the phenomenon, such as it is, greatly
misunderstood. Their direction, apart from the weather
conditions to which they are due, has no influence
whatever on the movements. Thus, if a migrant were
crossing the North Sea from Norway to Britain, one
fails to see why a north-west wind should not be as
suitable for the passage as a south-east, for both would
be beam-winds, and yet during the prevalence of the
former, such migration is practically at a standstill, and
during the latter at its climax.

High winds, from any quarter, are naturally
unfavourable, inasmuch as their very force may render
migration impossible or disastrous, or force the birds
out of their course when they encounter them. Strong-
flying species, such as Swans and Geese, do, indeed,
migrate under stress of weather that would be fatal to
most of the smaller species, and appear in our islands
during very unsettled periods. It is extraordinary,

PLATE VII.

Chart of North-Western Europe showing Typical Weather Conditions, with the Wind North-Easterly, over the British Isles

The region of highest pressure lies between the Hebrides and Iceland, with gentle gradients extending to and over the northern half of the British area and the southern and central portions of Scandinavia. The area of depression is centred to the south-east with a considerable gradient reaching across the southern half of the North Sea.

This type of weather is the most favourable of all for intermigration, in both spring and autumn, between the British area and Iceland; while it is also propitious for movements between our shores and those of Scandinavia at both seasons.

Such conditions are, however, unfavourable for movements to and from the British Isles and the countries lying to the south and south-east of them.

PLATE VII.

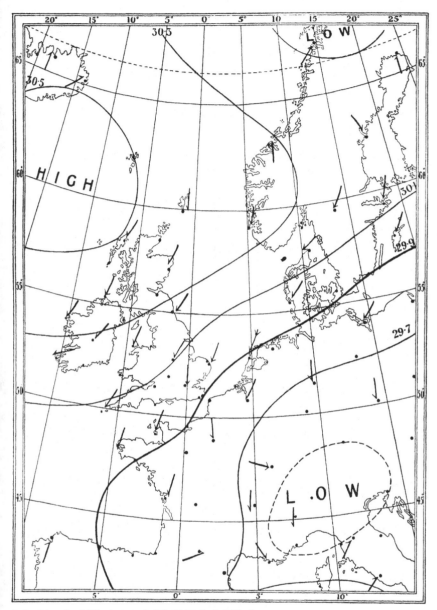

Reduced from Special Chart prepared at the Meteorological Office.]

NORTH-EASTERLY TYPE OF WEATHER.

Isobars are shown by black lines, with indications of the height of the barometer.

WIND.—Arrows flying with the wind, show direction and force, thus :—

⟶ = forces 5 to 7 (velocity, 28 to 40 miles per hour).

⟶ = forces 1 to 4 (velocity, 8 to 23 miles per hour).

however, what even small birds can brave, in the shape
of high winds, for a time, but a long flight against such
heavy odds is disastrous.   Blackbirds have appeared at
the Eddystone when the wind registered a velocity of
40 miles an hour.   Gales during the autumn drive
marine species, such as Phalaropes, Skuas, Petrels, etc.,
and even Gannets, on to our shores (some of them
occasionally in great numbers) and sometimes far
inland.   Later in the season, in winter, Guillemots,
Razorbills, Puffins, and Little Auks, are, in like manner,
blown ashore from their pelagic winter haunts, and
are not unfrequently carried far into the interior of the
country.

Temperature.—Temperature plays an important part
in both the spring and autumn movements.   In the
spring, a decided increase in warmth in the more
temperate winter retreats is an incentive to move
towards summer haunts ; and it has almost invariably
been found that the earlier arrivals of such migrants on
our shores are to be correlated with a rise of temperature
in countries to the south of the British Islands.   It is
worthy of note that, in not a few instances, such movements have been recorded for dates on which the
temperature of our islands was lower than immediately
before the birds appeared—a fact which clearly indicates
that the increase of warmth at the seat of emigration
was the incentive for the movements northwards.   This
rise in temperature in South-Western Europe sometimes
extends to the British area, prevailing over it to a greater
or lesser extent.

In the autumn, a decided fall in temperature warns
the summer guests that the time to seek the southern
winter retreats is at hand, and later compels the laggards

to migrate, through its effect upon their food supplies. Should the weather be unfavourable for a prolonged flight, the would-be emigrants have to await the advent of propitious conditions, and then a rush to the south or west, as the case may be, follows. Cold spells, however, are not unfrequently associated with anticyclonic periods, during which, too, the weather is calm, and hence suitable for departure movements from the area over which they prevail.

Fog.—It often happens that during an important migratory movement in the autumn fog prevails. On such occasions more birds than usual approach the lanterns of the light-stations, where they are sometimes killed in considerable numbers. This phenomenon is another effect of those anticyclonic conditions which have been alluded to as favourable for migration, and it is thus not surprising that the birds should encounter foggy weather during their great movements. There is also some direct evidence that birds lose themselves in foggy weather, since practically non-migratory species appear during its prevalence at unusual seasons at stations just off the coast.

Dr Shaw informs me that in the past few years he has had a large correspondence, chiefly through Dr Tressider, with various pigeon-flying associations in connection with pigeon-matches. This has led him to conclude that a successful pigeon-race requires anti-cyclonic conditions, so much so that the custodian of many thousands of pigeons sometimes waits in France until we signify the absence of cyclonic conditions. It appears to be cloud which is the obstacle to a pigeon's finding its way home, and possibly the same conditions may be operative in natural migration.

PLATE VIII.

CHART OF NORTH-WESTERN EUROPE SHOWING TYPICAL WEATHER
CONDITIONS, WITH SOUTH-WESTERLY WINDS, OVER THE BRITISH
ISLES

A depression lies to the north-west of our area, and embraces Iceland and
the Faroes ; while an anticyclone to the south of us covers the greater part
of France. Wind south-westerly over the British Isles and North-Western
Europe.

South-westerly weather is favourable for intermigration between England
and Southern and Central Europe, *i.e.*, for the arrival of summer visitors on
our southern shores in spring, and for the departure of emigrants from them
in autumn. It is also suitable for the passage movements (to the east in
spring and west in autumn) across the southern waters of the North Sea,
between the south-eastern shores of England and those of the Continent
lying in the same latitude. They present, however, weather barriers to the
performance of migration between Northern Europe and Iceland and the
British Isles.

PLATE VIII.

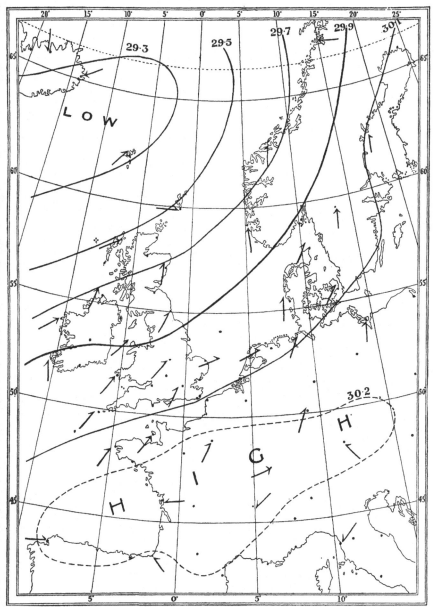

Reduced from Special Chart prepared at the Meteorological Office.]

SOUTH-WESTERLY TYPE OF WEATHER.

Isobars are shown by black lines, with indications of the height of the barometer.

WIND.—Arrows flying with the wind, show direction and force, thus :—

      —→ = forces 5 to 7 (velocity, 28 to 40 miles per hour).

      —⟶ = forces 1 to 4 (velocity, 8 to 23 miles per hour).

Hence it is worth while to consider the cloudiness of the various barometric distributions. Anticyclonic skies are not necessarily free from cloud, and in the winter there is a state of the sky which goes by the name of "anticyclonic gloom" among certain experts; but as a working rule in spring and autumn anticyclonic weather is generally fine.

Every migratory movement has its own particular and peculiar meteorological associations, for the conditions controlling them are often of a more or less complex nature. Most of them admit of explanation, when the official weather reports, which afford the key for their solution, are consulted.

The Weather and the Spring Immigratory Movements.—The meteorological phenomena which are associated with and influence these, must be sought in the weather conditions prevailing in those southern countries whence the migrants set out to reach the British Islands.

Birds often arrive in our islands at this season when the meteorological conditions with us are distinctly unpropitious, but a reference to the charts of the weather prevailing in the countries to the south of us invariably indicates that favourable weather prevailed there, and induced the birds to move northwards. The chief factor in these favourable conditions is an increase in warmth, the influence of which has already been treated of under the heading of Temperature. The southerly types of weather (see Plates V. and VIII.) are the most favourable for these spring movements across the Channel from the south.

The first arrivals of the summer birds appear, as a

rule, in March, and it may be remarked that the thermal peculiarities of the British area play an important part in determining their geographical distribution. The remarkable fact that the great majority of the summer visitors to our islands are first observed on the shores of the south-west of England and Ireland, has already been mentioned. This holds good even in genial seasons, but in cold ones it is almost entirely the case. Thus in March 1887, with its monotonously low temperatures, the arrival of six species, on twelve occasions, was recorded in the returns made to the British Association's Committee, *all* for the south-west. Again, in the cold March of 1885, every record but one of the fourteen chronicled was made in this same mild region of the British area. During the exceptionally cold and stormy March of 1883, only one species—the Wheatear—was observed on two occasions, both at stations on the west coast of Ireland, where the temperature was highest.

The Spring Emigratory Movements of birds which have spent the winter in our islands and are returning to their northern native homes, are influenced by the weather conditions which prevail in the British area. Here, it is found, other conditions being suitable for the sea passages to the north or east, that increase in temperature is again the main influencing factor, so that upon it depends, to a considerable degree, the extent of the departure movements. Thus it is not until April, and especially May, that the decided departures are recorded. In April the fine weather (anticyclonic) periods have varying emigrational values which depend entirely on their temperature. They are favourable if characterised by high, or moderately high, temperatures ;

or they may be distinctly unfavourable through being decidedly cold. Some emigration, of a straggling nature, it is true, is recorded during anticyclonic periods when moderately cold but calm weather prevails.

In spring, cyclonic periods, too, vary in their influences on emigration. They are, as a rule, unfavourable owing to their high winds and ungeniality. On the other hand, when they are of a mild type, and characterised by warm rain and soft breezes, following a cold anticyclonic spell in April, they favour to a northward movement from our islands.

The spring emigrations are embarked upon under precisely the same type of pressure distribution as that described as being favourable for the autumnal passage of birds across the North Sea to our islands—namely, when an anticyclonic area extends from north-western Europe to the shores of Britain. Under such meteorological conditions, the North Sea is spanned by fine weather, and moderate easterly to southerly breezes prevail. Some of these spring movements to the north are occasionally undertaken during somewhat unfavourable weather. Even in May there are records of emigration during sleet, cold rain, and north-east breezes, but it has to be explained that these conditions followed prolonged spells of inclement weather, and were genial in comparison with them.

Autumn Immigration.—The arrival movements of birds in the British Isles during the early autumn are undertaken under weather conditions which were favourable for their performance.

It is not until the latter half of September, and during October and early November, that the movements into our islands from the north-east are to be associated with the great weather changes of the autumn.

It has been ascertained then that *all* these great movements are due to weather conditions which have been already described as being favourable—namely, the prevalence of fine weather between Scandinavia and Britain. These conditions often follow the passing away from North-Western Europe of a cyclonic spell of a more or less pronounced nature, during the prevalence of which the ordinary course of the emigratory movements is either interrupted or rendered impossible. The anticyclone removes the cyclonic weather barrier, releases the pent-up migrants, and provides conditions favourable for migration, sometimes adding also an incentive in the form of a decided fall in temperature.

Autumn Emigration.—The chief feature in migration during the earlier autumn days is the departure of British summer birds, including those which have been described as partial migrants. July, in some seasons, has its ungenial spells, and these make themselves felt by our feathered guests, resulting in movements of a partial or a more decided nature. The influences inciting these incipient movements are a complete break-up of normal summer weather and the prevalence of unsettled conditions, not unfrequently accompanied by thunder and heavy rains and a decided fall in temperature. The result upon our summer visitants, or upon their young, on such occasions, is that many of them move from their accustomed haunts, and appear on the coast—sometimes at the lanterns—where the occurrence of those departing from our shores is duly chronicled. The species chiefly affected are the Thrush,

Redbreast, Wheatear, Whitethroat, Willow - Warbler, Swallow, Martin, Swift, and Cuckoo.

During August the regular departure movements of the autumn set in, and are usually performed under ordinary conditions—namely, fine weather. The influences, other than normal, are the same ungenial spells, especially if accompanied by cold, alluded to for July. These, however, are not frequent in most seasons, and yet few seasons are entirely free from them.

In September there are usually recorded some very decided movements, which may be fairly termed emigratory "rushes." These occur simultaneously with the advent of weather spells which, among other characters, are remarkable for a decided fall in temperature, sometimes amounting to many degrees. In one instance, on 15th September 1886, the difference in temperature amounted to as much as 20° in twenty-four hours, and naturally produced a marked effect on the emigration returns. The conditions causing such decided falls in the thermometer are chiefly northerly winds. Sometimes, however, these cold spells prevail with a light southerly wind, as was the case on 5th September 1885, when a cold, showery period caused much emigration. That low temperatures are the prime factors, is clearly demonstrated by the September records, inasmuch as there are unsettled periods which are not characterised by cold, whose influence on migration is comparatively insignificant.

The British autumnal emigratory movements of late September, October, and early November, are, in their meteorological aspects, identical with those which are associated with similar movements from Northern Europe, except that it is essential the conditions favourable for them prevail over the British area and to the southwards (see Weather Charts). Indeed, the movements are sometimes kept quite distinct from immigrations by the interposition of weather barriers to the north, which cut off migratory communication between our shores and those of Northern Europe. These barriers most frequently take the form of a subsidiary low-pressure area lying over the North Sea between Great Britain and Scandinavia.

The great emigrations from Britain and Ireland, like those from Northern Europe at the same season, set in on the passing, away of cyclonic conditions. The unfavourable conditions which have passed away have probably acted as a warning to many laggard migratory birds, while the cold adds an additional spur and swells the ranks of the departing birds.

During October local movements are observed, which are directly traceable to the influence exerted on emigration by a considerable lowering of the temperature over a particular area. Thus, for example, on 20th October 1883, there was a remarkable movement of Swallows to the south-east coast of Ireland. On this day there was a decided fall in temperature, the lowest readings being recorded for Ireland, where these laggard summer-birds had until then found congenial quarters. Again, on 10th October 1885, a movement to the southward of Thrushes and Blackbirds was recorded at stations in the north of Scotland, and in this instance, too, the meteorological data afford the information that a fall in temperature had occurred within that area.

The emigratory movements of winter are, as has been already stated (p. 37), attributable to the direct influence of severe weather conditions, in the shape of frost or heavy snow, and nothing more need be said regarding these simple weather influences on British bird-emigration.

Other references bearing directly upon the relations between meteorological and migrational phenomena will be found in the chapters dealing with Spring, Autumn, and Winter ; and also in those devoted to Bird-watching on the Eddystone and at the Kentish Knock Lightship : while the weather and its association with the movements observed for a whole year will be related in the study which treats of Fair Isle.

CHAPTER IX

THE MIGRATIONS OF THE SWALLOW, *HIRUNDO RUSTICA* [1]

THE various seasonal movements of the familiar Swallow afford an excellent type of those performed by the great majority of the smaller birds, which are Summer Visitors to Great Britain and Ireland, and whose breeding range also extends to higher latitudes on the continent of Europe, though not reaching to the extreme north, nor to Iceland.

In our islands the Swallow is, however, not merely a summer visitant, but also a bird of passage traversing our shores in spring and autumn on its way to or from its summer quarters in Northern and Western Europe. Its winter quarters are in Africa, chiefly to the south of the Great Desert.

Spring Immigration of Summer Visitants. — The records relating to the Swallow's return in spring are so numerous and complete as to enable one to speak with authority as to the date of the bird's successive arrivals on our shores, and also to trace with some degree of accuracy its gradual spreading over the country.

[1] The preparation of these complete and particular accounts has proved to be a most difficult undertaking. This arises from the fact that a number of the movements treated of are so intricately interwoven with, or so insensibly merge into one another, or are performed under such obscure conditions, as to render their discrimination and interpretation matters demanding most careful consideration.

There are several instances (which can only be regarded as phenomenal) of the appearances of Swallows in February. During March, however, a few annually appear on the south-west coast of England, in Ireland, the south-east of England, sometimes very early in the month, though these latter must be regarded as somewhat erratic visitors. There are also a few March records of their appearance in the south-eastern and south-western districts of Scotland.

It is not till April, however, that the vanguard of the spring hosts reaches our shores, and a careful analysis of dates shows that the average time of its appearance in different parts of our islands is as follows :—For south-western England, the beginning of the first week ; for Ireland, the end of that week ; for south-eastern England, early in the second week ; for south-western Scotland, the end of the same ; for south-eastern Scotland, the middle of the third week ; for northern Scotland, the fourth week ; and, lastly, it is not till the second week of May that the few swallows which are natives of Orkney reach their destination. These early immigrants are usually either single birds or pairs. Some ten or twelve days later than the arrival of this advanced guard, the appearance of Swallows in considerable numbers takes place, and they become gradually abundant throughout the kingdom. These initial hosts are followed by others, and so the influx proceeds during the rest of April and the first half of May, and beyond that date in the case of birds of passage en route for northern haunts beyond our shores. In backward seasons, such as that of 1887, when cold and unsettled weather with snow and sleet prevails, the vanguard may be delayed for about a week, but in that season its appearance was immediately followed by a "rush," and the birds became numerous and general only a little in arrear of their accustomed time.

In the Hebrides and north-western Scotland, the Swallow is not common, and is mostly observed on passage in small numbers. Occasionally it visits St Kilda in spring. It appears annually in Shetland on migration, chiefly about the middle of May and during June, but is somewhat irregular, both as to the date of its appearance and its numbers. In Ireland the immigrants continue to arrive in considerable numbers until about the middle of May, and in some seasons (1883, 1884, and 1886) as late as the third week of that month, but it is possible that some of these later birds are on passage on their way northwards.

It is evident from the statistics that the arrival of Swallows on the western sea-board is well in advance of their appearance further to the east. Not only is this so in the south of England, but even in Scotland the districts of "Solway" and "Clyde" almost invariably receive their swallows several days before the "Tweed" and "Forth."

The spring swallows are recorded as arriving on our southern shores chiefly during the daytime, mainly in pairs, but sometimes as many as six or seven together, and as flying low over the sea, the immigration lasting most of the day. They are also noted as coming in small parties, flock after flock, for several hours in succession, and are usually unaccompanied by any other kinds of birds. At the Eddystone, however, they have on several occasions been observed passing towards the Cornish coast during the hours of darkness, and with other migrants. Thus from midnight to 3 A.M. on 3rd and 4th May 1887, hundreds of birds, Swallows, and Wheatears, in company (as testified by the wings of the victims) with Reed-Warblers, Whitethroats, Wood- and Willow-Warblers, and Redstarts, were killed at the lighthouse. They also appeared at this same station during the early hours of 12th April 1902, along with many other species, and some were again killed against the lantern. Generally, however, few Swallows meet with disaster during their spring journeys, a very small number striking the lanterns, while fewer still seem to suffer from exhaustion.

Spring Passage to Northern Europe.—This movement of Swallows along our coast-line for their boreal homes, does not set in till the last days of April or early May, reaches its maximum about the middle of May, and may be prolonged till near mid June. Many of the earlier of these transient migrants reach our south coast in company with the Swallows that come to summer with us, but those which pour in during the latter half of May or in June are mostly travellers on their way to Scandinavia.[1] The main stream is confined to our eastern coast, and the North Sea is crossed by the main body of the migrants ere the northern limit of the mainland is reached, for only a comparatively small number seem to take Orkney or Shetland on their way to Northern Europe. A few Swallows yearly visit the Hebrides, including the remote Flannan Islands and St Kilda, during the last three weeks of May and early June, and it is possibly these birds, or some of them,

---

[1] According to the information of Professor Collett, the Swallow is seldom observed in Norway in April. In the first week of May examples appear singly, about the middle of that month more arrive, and between the 20th and 25th all, perhaps, have come.

and those which visit Sule Skerry and Foula, that find
their way to the Faroes,[1] and finally reach Northern
Europe by this far-western route. A few are also
observed about the same time on the north-west coast
of Scotland, probably en route for the north.

Autumn Emigration of British Summer Visitors.—
During the latter half of July, parties of Swallows are
recorded as visiting the island stations and lightships off
the east coast of Great Britain and the south-east of
Ireland. It may be doubted, however, if such appearances are of much significance, though it may be otherwise with some recorded during spells of cold weather.
But even if these were cases of real migration, it
may have been but partial, the birds merely seeking
better quarters within our area. On 19th July 1902,
however, a small party passed the Varne lightship, which
is stationed in the middle of the Channel and about
midway between Folkstone and Boulogne, at 11 A.M.,
flying from north-west to south-east.

It is not until the last week of August that Swallows
ordinarily begin to gather together prior to leaving
Scotland and the north of England. Then there is
a decided movement southward, and, along with Wheatears, Redstarts, Sedge-Warblers, Willow-Warblers, and
Tree-Pipits, they are observed at various stations, both
on the coast and inland, and some even quit our islands.
There is no evidence, however, that these birds leave the
country in any but very small numbers, and most of them
probably tarry for some time in the south of England
before crossing the Channel. The Irish movements in
August are less pronounced, but the returns show a

[1] Herr Knud Andersen informs me that the Swallow appears not
uncommonly, as a straggler, in the Faroes in May.

decided increase of visitors to the coast stations, and indicate the setting in of autumn emigration.

In September the southern movement becomes
general throughout the whole country, and reaches its
maximum between the middle and end of the month.
During its early days there is the first evidence
of pronounced departure from our shores, and the
cross - Channel emigration then proceeds throughout
the autumn. At the Eddystone during the latter
half of September and first half of October 1901, I
observed considerable numbers passing southwards.
They were usually in parties of a dozen or more, which
comprised old and young birds. All the movements
were timed between 7 and 11.30 A.M.

The beginning of October shows a decided falling
off in the numbers departing from the northern districts,
especially in the west ; but the southward movement is
well maintained during the first half of the month from
the east and south-west of England and the south-east
of Ireland. By the middle of the month the emigration
from Scotland and the north of England is over, and
Swallows observed after that time on the east coast
of Britain are migrants from Scandinavia, which
since September have been passing along that coast,
mingling with our own departing birds, so that in many
cases the two movements are indistinguishable. After
the middle of October a considerable diminution is
observable, except on the coast of the Channel, where
the efflux is maintained throughout the month.

During the first half of November stragglers are still
to be seen on the east coast of Great Britain and the
south-east of Ireland, but there are very few records of
observations in the west of Scotland, and not many

from the north-west of England. From the south of England many departures occur annually till the middle of the month, while stragglers are to be seen later, especially in the south-west. December Swallows are *raræ aves*, and were only observed in one year of the British Association inquiry. The autumn of 1880 was remarkable for the protracted stay of the Hirundinidæ, and a few belated Swallows were recorded on the south coast of England in the last week of November, while in December one was observed at Bournemouth on the 7th, and two at Eastbourne, and one at Woolmer on the 11th, the weather until that time having been mild.[1] Others were observed in the Decembers of 1891, 1894 (a number), and 1896.

Autumn Passage along the British Coast from Northern Europe.—The return of the Swallows which have summered in Scandinavia (accompanied by their young), and their passage along our coasts usually takes place from the middle of September[2] onwards, the 9th (in 1884) being the earliest day on which their movement is recorded. The passage is well maintained during the rest of the month, and is prolonged on the part of small numbers to the first or even second week of October. Some of these travellers from the north are, perhaps, induced by our milder climate to tarry, and it is possibly such laggards that occur on or near our east coast in November, and account for the late-

[1] Mr Joseph Agnew, light-keeper, states that a Swallow was caught on the Monach Isles (with the exception of St Kilda, the outermost of the Hebrides) in January 1887, but he unfortunately furnished no further particulars of the occurrence.

[2] Professor Collet states that Swallows begin to leave southern Norway the first week of September, and that he has known individuals to remain there so late as the middle of October.

ness of migration there as compared with that on the west coast.

In Shetland and Orkney there are only slight and irregular appearances of these returning Swallows of passage, and but feeble evidence of their taking the Hebrides on their return journey, though the records indicate such a transit during September and the first days of October. There are passage movements on the part of Irish birds discernible in the south-west of England to the third week of October, with occasional stragglers to the middle of November.

It has been already remarked that after their arrival on our shores, Swallows on autumn passage mix with our native birds then emigrating, and it is no longer possible to trace the former, though they doubtless form the bulk of the rearguard movements of the autumn.

Autumn Passage from Western Europe by East-to-West Route.—During my sojourn in the Kentish Knock lightship in the autumn of 1903, I was much interested to observe Swallows coming from the east and flying towards the coasts of Essex and east Kent. Such flights were witnessed on seven occasions between 26th September and 16th October. These migrants from the opposite shores of the North Sea passed the ship in parties composed of young and old, and were timed between 8 A.M. and 2 P.M. Spring passages, in the reverse direction, do not appear to have come under notice.

In September Swallows are recorded at the light-ships off the mouth of the Thames and the Kentish coast as coming from the south-east, occasionally in considerable numbers. This would seem to imply that they were leaving the French or Belgian coasts, perhaps to sojourn in England before finally departing for their southern winter-retreats.

Further Observations on the Autumn Movements.— At the best stations for observing the emigration of the Swallows from our shores, it usually takes the form of the continuous passage of small parties not exceeding a score, and as this may last for hours, vast numbers thus depart.   They have, however, been observed to assemble on the south coast in thousands and fly away *en masse*, but it is only occasionally such departures are recorded.   The earliest troops to cross the Channel are composed of old and young birds.   It has, however, been noticed that the large congregations at various points on the south coast, whether preparing to emigrate, or in actual movement, consist in some instances chiefly or entirely of young birds, and in others wholly of adults. More frequently, however, the number of old birds is in normal proportion to that of the young.   Swallows are frequently seen emigrating in company with House-Martins and occasionally with Sand-Martins.

The flittings-away of the Swallows which have summered in the British Isles, and of those which visit our shores as birds of passage proceeding southwards, are mainly undertaken during the daytime.   On the south coast some of the great movements are recorded as in progress from early morning until noon, others proceeding until night sets in.[1]   There are records, however, of night movements.   Thus, at the Casquets, west of Alderney, on 1st October 1880, Swallows, with

[1] At the Nab lightship, 1st October 1886, Swallows were recorded as passing south at intervals, twenty at a time, from dawn to dark.   The returns from Hanois lighthouse, on the west coast of Guernsey, show that Swallows pass southward from 6 A.M. to 8 P.M.

other birds (Song-Thrushes, Ring-Ouzels, Land- and Water-Rails, and a Woodcock), occurred from 11 P.M. to 3 A.M. Two hundred Swallows struck the lantern.

During the autumn (and also spring) migration, the English Channel is probably crossed by many routes, but there are certain much-used points of departure, to reach which the birds shape their course. Beginning in the west, we find among them the Land's End, the Lizard, the Eddystone, Start Point, Portland Bill, Isle of Wight (St Catherine's Point and Nab light-vessel), Beachy Head, and Dungeness. On the Dorset and Hampshire coasts, Swallows are recorded as proceeding to the eastward in the autumn. In Sussex, too, the flight is easterly, towards Beachy Head, just before arriving at which many birds cross the Channel.[1] Others still pursue their easterly flight, and finally cross the Straits of Dover. There may be other routes taken, but the points of departure just named are those which result from the investigations with which I have been associated. There are, however, some records of Swallows occasionally moving westward along the south coast, perhaps a continuation of the movements from the east across the southern waters of the North Sea. Thus a cross-movement of departing birds then occurs. The coast-line is closely followed by many of the Swallows moving south, more especially by the birds of passage.

Among the birds performing similar migrations as summer visitors to the British Isles and North-Western Europe (due allowance being made for differences in the

[1] When crossing the Channel between Newhaven and Dieppe, during the daytime, in September, I have seen Swallows passing in a south-easterly direction towards the French coast.

dates on which they are performed) are : the Ring-Ouzel, Wheatear, Whinchat, Redstart, Common and Lesser Whitethroats, Blackcap, Garden-Warbler, Chiff-chaff, Willow - Warbler, Sedge - Warbler, Tree - Pipit, Red - backed Shrike, Pied and Spotted Flycatchers, House- and Sand-Martins, Swift, Nightjar, Wryneck, Cuckoo, Corn-Crake, Common Sandpiper, Whimbrel, etc. The Wheatear and the Whimbrel proceed to Iceland and the Faroes, as well as to North-Western Europe.

CHAPTER X

THE migrations of the Fieldfare have been mainly selected as being well suited to illustrate the various movements observed in the British Islands of an important and numerous class of migrants—namely, the Winter Visitors.

The British migrations of this species are those of (1) a winter visitor to our islands from North-Western Europe, and (2) of a bird of passage in autumn and spring, when en route between its northern summer-home and winter-quarters in Southern Europe and Northern Africa. In addition, British winter movements (due to the pressure of climatic conditions), including emigration beyond our shores, are annually performed.

The home of the Fieldfares which visit the British Islands is in Northern Europe, presumably Norway for the most part. The species does not breed in Iceland, as the Redwing does, and there is no evidence to show that any of the small colonies established in various parts of Central Europe (Pomerania, Thuringia, and Bavaria) ever contribute to the throng that arrives on our shores in the autumn.

Autumn Arrival in British Isles.—The Fieldfare seldom quits its boreal summer haunts until October.

There are, however, many records of the appearance in Great Britain of odd birds, and, more rarely, small parties in September, but such occurrences must be regarded as somewhat unusual.[1]

There are annual arrivals of comparatively small numbers in the first half of October; but at Fair Isle in 1907, a number arrived on 2nd October, and very many on the 10th. It is not, in most years, however, until after the middle of the month that the first of the great autumnal immigrations is to be expected; for the date of the northern exodus is dependent upon the nature of the season, and especially on the luxuriance of the crop of berries, in Scandinavia, and, as a rule, the birds do not move southwards until the third or fourth week of October. They continue to arrive on our coasts in considerable numbers until mid-November, the 19th being the latest date for the years covered by the inquiry instituted by the British Association, and the 24th for Fair Isle.

The following are the dates of the chief immigrations recorded during the years 1880-87, and at Fair Isle for 1906, 1907, 1908, 1910, and 1911 :—

1880. October 21-28.   November 18.
1881. October 19.
1882. October 15-16, 18-19.
1883. October 19, 28-30.   November 1-2, 8.
1884. October 24, 29.   November 2, 4, 12.
1885. October 14, 31.   November 8, 10-12.
1886. October 28, 29.
1887. October 26.

---

[1] The most remarkable of these early immigrations was the occurrence of a large flock near Norwich on 9th September 1880 (T. Southwell).

1906. October 22, 23, 26.   November 3, 22.
1907. October 10, 12, 18.
1908. October 24, 31.   November 24.
1910. October 21, 24-29.
1911. October 14, 21, 22.

It will be observed that in most seasons the birds arrived in a series of pronounced movements, while in others a single "rush" only is chronicled. When the latter is the case, it must be remarked that it was preceded or followed (or both) by a steady influx covering the ordinary period of the autumnal incoming (7th October to 22nd November). On many occasions these great immigrations cover much of the eastern sea-board, which is chiefly affected from Unst, in the Shetland Isles, to, or perhaps beyond, the Humber.

After arrival the immigrants quickly find their way to accustomed winter quarters in the British area, including those in the western districts, which are largely, but not entirely, reached by an overland flight from the north-east and east.

A migratory stream of Fieldfares, though one of much less extent, reaches the west coast of Scotland, where it is chiefly observed at the inner isles, but extends as far to the west as St Kilda, the Flannan and the Monach groups, and comes much under observation at the rock stations of Skerryvore and Dhu Hearteach. The Outer Hebridean branch of this stream reaches the north coast of Ireland, whence numbers of the birds proceed inland to winter quarters. Regarding these western movements, it must be observed that it seems likely that a small proportion of the Fieldfares regularly travelling southwards *via* the Outer Hebrides may reach that far-western route by way of the Faroes, which islands are visited annually in the autumn [1]—an interesting fact, since the bird does not summer in Iceland, and, moreover, one which indicates an astonishing extension westwards of the right wing of the hosts moving southwards on the approach of winter.

Autumn Passage and Emigration. — The autumn passage from the northern summer haunts to winter quarters beyond the British Isles is chiefly observed on our east coast during the latter half of October and the first half of November—the two sets of migratory Fieldfares (the British and South-Western European winter visitors) doubtless arriving in company. Thus many Fieldfares quit our southern shores very shortly after their arrival, and consequently the dates of immigration and emigration closely correspond. A number of the migrants observed on the west coast also proceed southwards; some of them along the east coast of Ireland, and thence across St George's Channel; others by way of the west coast of England and Wales. These birds finally quit our shores at points on the western section of the south coast of England, particularly between the Eddystone and the Scilly Isles.

Winter Movements and Emigration.—On the advent of snow and cold, the Fieldfares quit the higher grounds, which form their usual winter quarters in our islands, and seek the lowlands, the coast,[2] and the islands lying off it.

---

[1] Mr Knud Andersen informs me that some Fieldfares occur on migration in both spring and autumn at the Faroes.

[2] The appearance of this bird in numbers on the coast in winter has led some observers to suppose that a renewal of the immigratory movements from Northern Europe has occurred, whereas it is directly associated with and is the result of the weather conditions prevailing in our islands, which have driven the wintering Fieldfares from the inland districts.

In seasons of exceptional cold and heavy snow, vast numbers pass southwards along our coast-lines and overland, en route for the southern counties, while many cross the Channel for South-West Europe. Many, too, especially after or during snow, are observed passing westwards along the south coast of England and its vicinity, in company with Thrushes, Redwings, Blackbirds, Starlings, Larks, Linnets, Lapwings, etc., in search of the milder conditions usually to be found in Devon, Cornwall, and the Scilly Isles. Emigrants from the mainland of northern Britain then visit the Hebrides; and numbers enter Ireland from Scotland and north Wales; but none of the numerous Fieldfares which sweep along the south coast of England appear to seek the Sister Isle from the south-east by a passage across St George's Channel, as do Song-Thrushes, Starlings, Larks, and other refugee British species. In Ireland, during severe periods, many leave their ordinary winter haunts and pass southwards and westwards for the milder areas to be found in the vicinity of the Atlantic or to quit the country.

The time at which these winter movements take place varies according to the nature of the season. In 1886 great cold set in as early as 22nd November, and was the cause of much migration, and an exodus from the southern shores of Britain, which continued until the 26th. The movements may be local or general, and if a series of cold snaps occurs, a corresponding series of spasmodic migrations results; but should the storm be widespread, general emigrations follow.

A small number of Fieldfares winter in southern Scandinavia, and in exceptionally severe seasons some of these are driven southwards and westwards, and this, no doubt, accounts for the occasional appearance of small numbers in the depth of winter in Shetland and Orkney and at the Faroes (Andersen).

Spring Immigration from the South.—Towards the end of March, the Fieldfares which have wintered in countries south of the British Islands (including, no doubt, the birds driven from Britain by the severities of the past winter) make their appearance on our southern shores, which form an important stage in their spring journey northwards. These return passages across the Channel are continued at intervals throughout April, and are sometimes observed down to the early days of May. During these movements the birds are recorded as arriving at night or in the earliest hours of the morning, and are usually accompanied by Redwings, Thrushes, Blackbirds, Starlings, Wheatears, and other species.

The immigrants which arrive in England during March do not appear to move northwards at once, but sojourn with us for a little time before departing for their summer quarters in Northern Europe.

Spring Passage and Emigration.—The departure from our isles of the Fieldfares which have wintered with us, and of the birds of passage also on their way to the northern breeding haunts, does not commence until the latter days of March (22nd earliest) or the early days of April, and ordinarily lasts until the second week in May, but in some seasons is prolonged until about the middle of that month, while stragglers are annually observed as late as its last week.

Some of the great spring emigrations cover considerable sections of the eastern sea-board, having been observed from the Farne Islands to Orkney and Shetland, and from the Wash to the Firth of Forth.

Departure movements are also witnessed on the west coast of Britain. These are somewhat feebly marked at the English, Welsh, and Manx stations, but become more pronounced on the Scottish coast (including the Hebrides), which receives the Irish emigrants en route for the north. There is also much overland migration from the western districts to the east coast, for embarkation, performed throughout Great Britain.

Before proceeding to the coast for departure, Fieldfares assemble and form flocks in the various districts in which they have wintered, and are then very noisy and restless for several days before they finally take leave of their winter haunts.

The earliest emigrants quit our shores chiefly in small parties, either alone or at the same time as Rooks, Starlings, Bramblings, Mealy Redpolls, Yellow Buntings, Skylarks, Meadow-Pipits, Song-Thrushes, Redwings, Ring-Ouzels, Wheatears, Redbreasts, Hedge-Accentors, and Woodcocks. Those that follow in the latter half of April and in May are observed mainly on the east coast and at the northern islands of Orkney and Shetland, in great flocks, and in company with many other migrants—Blackbirds, Whinchats, Redstarts, Blackcaps, Common and Lesser Whitethroats, Willow-Warblers, Pied Flycatchers, Swallows, Corn-Crakes, Whimbrels, and other species, including most of those already named as observed during the earlier movements.

General Remarks.—This species is extremely wary, and is less frequently killed or captured at the lanterns of the light-stations than any of its congeners.

Among the winter visitors to the British Isles performing migrations similar to those of the Fieldfare (due allowance being made for the variation in the dates of their comings and goings, which in most cases are not material) are: the * Redwing, Brambling, * Mealy Redpoll, Grey Plover, *Turnstone, Great Snipe, Jack Snipe, *Knot, *Sanderling, Bar-tailed Godwit, and a number of aquatic species (Ducks, Geese, Grebes). Those marked thus * also come to us from Iceland, as well as from the north-east.

Many individuals of species which are resident in the British Isles, such as the Song-Thrush, Blackbird, Redbreast, Hedge-Accentor, Chaffinch, Siskin, Short-eared Owl, Merlin, Water-Rail, Golden Plover, Woodcock, Common Snipe, etc., are summer visitors to northern countries, and perform identical migrations as winter visitors to, and birds of passage on our shores.

CHAPTER XI

THE MIGRATIONS OF THE WHITE WAGTAIL, *MOTACILLA ALBA*

THE migrations of the White Wagtail are those of a Bird of Passage traversing our shores when en route between its summer quarters in North-Western Europe, the Faroes, and Iceland, and its winter retreats to the south of us, which extend to Northern Africa.

Its movements present several points of interest—among others, the fact that it is one of the species, not many in number, whose main lines of migration in the British Isles lie along our western sea-board and its islands, and not on the east coast.

There are records of the birds' nesting occasionally in various parts of Great Britain.

Spring Immigration and Passage Northwards.—The first White Wagtails to arrive from the southern countries in which the winter has been passed, have been detected on the south coast of England as early as the latter half of March.[1] It is not, however, until the early days of April that their appearance may be generally expected, and the scene of their arrival is chiefly on the western section of our southern coast-line.

[1] In 1872 it is said to have occurred at Plymouth on 3rd March, which is the earliest date known to me for its appearance.

Their in-comings in small numbers continue until the middle of the month, after which there is an increase, which is maintained until mid-May, after which date the immigration practically ceases.

After their arrival on our southern sea-board, some of the immigrants tarry for a while before resuming their journey.   Others soon move northwards, the great majority proceeding by the western routes, where there are instances of the bird's appearance, even in the Clyde area, soon after mid-March.   With April the general advance northwards sets in, and proceeds until about the middle of May ; but the rearguard is observed on its march northwards at the Shetland Isles, down to the latter days of the month.

On the west coast, we are able to trace the passage from the Scilly Isles and Cornwall to the Welsh coast, the Isle of Man, the Solway and Clyde areas, and hence coastwise to Cape Wrath, the north-western limit of the mainland of Great Britain.   In Scotland two more western routes are also followed, the outermost of which traverses the Hebrides from Barra Head, their southern extremity, to the Butt of Lewis, at their northern apex, while a number of the migrants visit the still more outlying Monach, St Kilda, and Flannan groups of islands, and North Rona.   Another line of flight is *via* Dhu Hearteach, Skerryvore, and the inner islands, and comes especially under notice at Tiree and Coll.

To turn now to the Irish coasts, we find that there are only a few records for the spring passage for the east coast of Ireland, but it will indeed be a very surprising fact if the bird is not on regular passage along that coast-line.   The west coast of Ireland is, however, regularly visited, and probably the birds traverse the Outer Hebrides on quitting the Irish coast.

As compared with the migrants proceeding north-wards by the western routes named, those which move along the eastern sea-board are but a feeble troop. This contrast with the west coast may, perhaps, be accounted for by the fact that the bird is an abundant summer visitor to the Faroes and Iceland, and that those seeking Scandinavia also chiefly follow the western route. Be this as it may, it seems to be well established by observation that only a small number arrive on the eastern section of the south coast for passage along the British shores of the North Sea, where a few are observed up to the north-eastern limit of the mainland.

The great numbers of these birds on passage mov-ing along our coast-lines in spring finally reach the Shetlands. There the White Wagtail is one of the commonest birds of passage, and these northern isles seem to be the main scene of its departure from the British area in spring, as it is of its arrival in autumn. It occurs annually in the Orkneys in spring, and is regularly observed in the most western and southern isles of the group—namely, at Sule Skerry and at the Pentland Skerries. The Orkneys and Shetlands too, no doubt, mark the parting of the ways, some of the migrants proceeding to the north-west for Faroe and Iceland, others to the north-east for Scandinavia. Some of the birds travelling along the westernmost route may, after traversing the Hebridean islands, proceed direct to the Faroes, and thence to Iceland.

As to the dates at which these spring movements are performed, the records from Fair Isle, where the bird is very common, furnish much reliable information,

I.   O

the result of careful observations made during six successive seasons (1906-1911). The earliest date for its appearance is 5th April, and the period during which it is most abundant is the fortnight which covers the last week of April and the first week of May. After this, stragglers only appear in small parties, and the latest date on which its passage is recorded is 9th June.

Autumn Passage South and Emigration. — The autumn movements from the summer homes towards the winter retreats follow the same lines of flight along our shores as those already indicated for the spring, save, of course, that they are in a reverse direction. The birds are then more abundant than in the spring, since the ranks of the migrants are recruited by the young birds a few weeks old. It is usually seen in parties of from ten to twenty individuals, but I have seen as many as fifty together at Fair Isle and thirty at St Kilda. These migrants are generally actively engaged in the capture of flies, which are very numerous among the rotting seaweed which fringes the beaches at high-water mark; they are numerous, too, in the neighbourhood of stations at which fish is exposed for drying, where also dipterous insects abound.

The earliest date for the bird's appearance at this season at Fair Isle is 9th August 1909. It becomes more numerous as the month advances, and the movements continue in full swing until mid-September. After this the passage, though still in progress, is participated in by fewer individuals, but small numbers are observed annually until mid-October. Odd birds, however, occur in November, and one obtained on the 9th of that month, in 1908, was forwarded to me for

my satisfaction. There appear to be few records of the
White Wagtail's departure from the southern shores
of England, but during my residence in the Eddystone
lighthouse in the autumn of 1901, I observed it crossing
the Channel towards the coast of France on several
occasions between 31st September and 14th October.
I only detected one of these travellers amongst the
Wagtails that migrated during the daytime; but birds
of this species took part in some of the great night
movements, and were captured at the lantern from 2
to 4.30 A.M., among crowds of other emigrants leaving
the English shores. At both the Flannan and Fair
Isle lighthouses, I have on several occasions seen them
at the lanterns during the small hours of the morning,
chiefly in company with Wheatears.

On the south-eastern and eastern sections of our
coastline the White Wagtail is, I believe, overlooked,
and we have yet much to learn concerning the details of
its distribution there during migration at both seasons.
Recent investigations, however, especially those of the
Misses Baxter and Rintoul, have added considerably
to our knowledge, and we are able to trace the movements from the east coast of Ross-shire to the southern
shores of the Firth of Forth.

CHAPTER XII

THE migrations of the Song-Thrush, though of a very varied nature as a whole, have been selected as being in the main typical of those performed by the group of British migratory birds which have been defined as Partial Migrants — species some of the individuals of which remain throughout the year in our islands, while others only spend the summer with us.

In addition to being a resident and a summer visitor, this favourite songster is also a winter visitor to the British area, a bird of passage along our shores in the spring and autumn, and lastly, a winter migrant—a fugitive evicted from its usual haunts by severe weather. Indeed, the Song-Thrush furnishes us with a most excellent example of the complex nature of the phenomena of bird-migration as observed in Great Britain and Ireland; and its various movements cover nearly the whole year.

Autumn Emigration of British Summer Visitors.— Though this Thrush is a permanent resident in certain districts, more especially in the gardens and immediate neighbourhood of cities and towns, where even in Scotland a considerable number remain throughout the year, such residents form only a moiety of our native birds of this species. At the end of summer and in the early autumn, a considerable number of the Thrushes which have reared their broods with us, especially those which inhabit the elevated and northern districts, emigrate towards the south.[1]   There are a few records for July which undoubtedly relate to migration of a partial nature.   These, however, must be regarded as exceptional and due to pressure of very unsettled weather, especially thunderstorms.[2]

During the second week in August, but chiefly towards the end of the month, there are clear indications that emigration is taking place.   The birds then come under notice on the coast; while the light-stations afford positive evidence that Thrushes are quietly slipping away from Britain.[3]   Though there are no marked movements recorded for this month, yet there is unmistakable evidence that a gradual and steady emigration is in progress on all the coasts of Britain.

These flittings-away become more general and pronounced in September, and are witnessed throughout October; but the movements then are often of a very complex nature, and are difficult to interpret, since they

[1] Mr T. G. Laidlaw, whose home in Peeblesshire lies 900 feet above the sea, informs me that the thrushes leave that district "to a bird" in the autumn, and return during the early months of the year.   A young bird ringed in a nest in Aberdeenshire on 4th June 1910, was shot in Portugal on 6th November.

[2] On 8th July 1882, five Thrushes struck the lantern at Slyne Head lighthouse (west coast of Ireland), one of which was killed.   In 1885, on 3rd and 11th July, several thrushes are recorded at the inner Farne.   On 10th July 1905, one visited the lantern of the Bell Rock lighthouse. A few were observed at the Eddystone lighthouse on 26th July 1909.

[3] As early as 1st August 1884, six Thrushes struck the lantern of Dhuhearteach Rock lighthouse, two being killed ; while at the Eddystone, on 8th August 1902, Thrushes were crossing the Channel along with Wheatears, "Warblers," and Curlews, and visited the lantern from 10.30 P.M. to 1 A.M. (9th).   Ringed British birds have been recovered in France and Portugal.

I.   O 2

become inextricably mixed with those of the birds of passage from Northern Europe.

During these months, especially in September, the Thrush departs in company with various species which have spent the summer in our islands, and its emigrations are recorded from all sections of the British coast and from the east and south coasts of Ireland, always during the hours of darkness. The Thrush is, however, emigratory to a lesser degree in the Sister Isle, owing to its having a milder climate than that of Great Britain.

Late in the year, the emigratory movements, which doubtless include many of the recently arrived winter visitors from Northern, and perhaps Central Europe, are dependent on and synchronous with more or less severe weather conditions, and these will be duly treated of under Winter Movements.

Autumn Immigration and Passage.—There is no evidence whatever of the arrival of the Thrush upon our shores as winter visitor and bird of passage from North-Western Europe until the third or fourth week of September, when it appears with great regularity, in company with the early Redwings, Bramblings, Siskins, Redbreasts, Goldcrests, and sometimes Woodcock, Jack Snipe, and Short-eared Owls.[1]

The immigrations continue during October and to the end of the third week of November, or a few days beyond. During this period there are "rushes" of a more or less pronounced nature to our shores, when for several successive nights Thrushes pour in upon our northern and eastern sea-boards in vast numbers.

[1] Prof. Collett, *Oversigt af Christiania Omegns ornithologiske Fauna*, p. 27, says that the Thrush departs from the Christiania district during September, and continues to do so until the first days of November.

These immigratory movements are observed on the east coast of Britain, from the Shetlands to Norfolk. The Thrush is also abundant on the west coast, including Ireland, and visits such outlying stations as Sule Skerry, the Flannans, the Monach Isles, St Kilda, and the isolated rocks of Skerryvore and Dhu Hearteach.

The Thrush's travelling companions, in addition to the species already mentioned, are chiefly the Fieldfare, Ring-Ouzel, and Blackbird. Along with these species, many Thrushes perish at the lanterns of the light-houses and light-vessels, especially when the night is hazy or there is light rain.

The majority of the winter visitors enter Ireland, however, at the south-east corner, where immense numbers arrive during October and November from south-western England, many of them after passage along the south coast.

Many of the immigrants upon arrival, or soon after, proceed as Birds of Passage along our eastern, western, and southern coasts, and finally quit Britain, the majority to seek more southern lands, others to cut across St George's Channel in order to winter in Ireland. Others, again, remain as Winter Visitors, and work their way to inland quarters by overland journeys. Numbers of these birds, however, quit our islands, after a longer or shorter sojourn, under the pressure of severe weather conditions.

Autumn Immigration from Western Europe. — Investigations at the Kentish Knock lightship, in the autumn of 1903, lead me to conclude that Song-Thrushes cross the southern waters of the North Sea by an east-to-west route from the coast of Holland to that of south-eastern England. During my sojourn on that favourably situated station in the North Sea, I, on several occasions between 18th September and 19th October, observed and captured Song-Thrushes at the lantern between the hours of midnight and 4 A.M. On each of these occasions they occurred with such species as Redstarts, Pied Flycatchers, Mistle - Thrushes, Blackbirds, Meadow-Pipits, Wheatears, Starlings, Sky-larks, Chaffinches, Rooks, and Jackdaws. All these may have been of Central European origin, and it is most significant that I *never* saw this bird in company with any of the species with which the northern Thrushes invariably arrive on our shores—a fact which has led me to conclude that the visitors at the Kentish Knock arrived there from the east. As yet, I have no information relating to the return spring passage of the Song-Thrush, nor of some other species which certainly traverse this line of flight in the autumn, and doubtless do so in the spring also.

Winter Movements.—The great emigratory movements of the winter sometimes commence in November,[1] and are continued during December, January, and February.[2] They are synchronous with outbursts of cold, of snow, or of extremely unsettled weather. Such untoward conditions may prevail generally over our islands, or they may be circumscribed ; and their influence on the emigrations of the Thrush is in direct consonance with the distribution of the disturbing elements.

[1] In 1886, as early as 4th and 6th October, there were great emigratory movements on all our coasts, due to extremely unsettled weather, with thunder in the north and north-west, accompanied on the 5th by a great change of temperature—a fall of fifteen degrees below that of the previous day.

[2] There are also movements during March in some years ; but they are of a local nature, and are not to be regarded as emigratory.

During the period in which ordinary climatic conditions prevail, little or nothing is recorded. In others the few local movements are traceable to topical weather conditions. But sooner or later during each season great outpourings take place, often extending over several successive days and nights and affecting all our coasts. The Thrushes affected are not merely our would-be resident birds, but a very large proportion of them are, no doubt, the immigrants lately arrived from the north, which, as winter visitors to our islands, remain until compelled to move further south or west.

The first move on these occasions is to the coast, especially the west coast and its off-lying islands, where some tarry, and even remain to perish. Others pass along both the east and west coasts of Great Britain; many of those following the former route sweeping along the south coast westward, and crossing the channels for the Continent and Ireland. Many, too, seek Ireland from the north and east. Emigrations from the Sister Isle are also recorded during winters of exceptional cold.

Should the cold spell be of great severity, or be seriously prolonged and widespread, many perish even in such usually safe retreats as the Scilly Isles, and at Valentia, or other isles off the west coast of Ireland, which are largely sought on such occasions. No doubt, too, many of the emigrants perish in their Continental asylums, for after winters of almost arctic severity, such as that of 1880-81,[1] the Thrush was conspicuous by its absence, or by its scarcity, in most districts in our islands.

[1] During this winter twenty days of hard frost and sixteen days of deep snow prevailed on the west coast of Ireland. It was much more severe elsewhere. See Chapter VII. for further instances of severe seasons.

Spring Immigration of Summer Visitors.—Among the records relating to the movements of this species during February, there are many which clearly indicate that the Thrushes which left us in the autumn to winter in countries to the south of us are now commencing their return to our islands for the spring and summer. Thus on 19th February 1903, at the Eddystone, several Song-Thrushes, with Mistle-Thrushes, many Starlings, and other small birds appeared at 9.30 P.M. On the following night, at the same station, Song-Thrushes again appeared with Mistle-Thrushes, Starlings, and Lapwings. These immigrations are performed by small parties during mild periods of the month, and are chiefly observed on the southern coasts of England and Ireland.

Such return movements are continued during the first half of March, when arriving Thrushes, in company with Blackbirds, Larks, Meadow-Pipits, Starlings, Lapwings, and Curlews, are recorded from the south coast of England northwards to the western isles of Scotland, and from the south and south-east coasts of Ireland.

The arrivals on the south coast of England take place during the night or early morning. In Ireland they are recorded both for the hours of darkness and during the daytime, and the birds are noted as proceeding in a north-westerly direction at the south-east stations, where the movements come chiefly under notice.

In most instances the return is a gradual one, performed by small companies and at intervals, but occasionally in March in "rushes" with the other species already mentioned.

There are also migrations from various British haunts, where the winter has been spent, to summer quarters in our isles. These usually commence in February, but March is the main month during which these local movements of home-bred Thrushes are undertaken.

Spring Emigration of Winter Visitors.—About the middle of March the Thrushes which have wintered in Tiree and other islands off the western coasts of Scotland and in Ireland are recorded as taking their departure for the mainland.

It is not, however, until the latter days of March[1] and the early part of April that the birds which have wintered in our islands leave our shores to return to their summer haunts in Northern and perhaps Central Europe.[2] The emigrations proceed throughout April, when large numbers are observed; a few are seen during the early days of May; and stragglers have been detected in rushes of migrants at the beginning of June.[3]

These emigratory movements come most under notice on the north-east coast of England, the east coast of Scotland, and at the Orkney and Shetland Islands. They are also in evidence on the west coast of Britain, in Ireland, and at the Hebridean Islands.

---

[1] On 22nd March 1909, a few appeared in a rush of migrants at Fair Isle. The other species observed were: Fieldfares, Redwings, Blackbirds, Chaffinches, Yellow Buntings, Pied Wagtails, Skylarks, Starlings, Rooks, Golden Plovers, and Lapwings.

[2] Professor Collett, *Oversigt af Christiania Omegns ornithologiske Fauna*, p. 26, gives from the early to the last days of April as the period for the Thrush's arrival in spring in the Christiania district.

[3] At Fair Isle, on 8th June 1907, Thrushes occurred in company with Wheatears, Blackcaps, Garden-Warblers, Whitethroats, Swallows, House- and Sand-Martins, and a Snow Bunting.

With April, perhaps earlier, the British emigratory movements doubtless become merged with those, to be dealt with next, of the Thrushes which are on passage along our coastline, proceeding from their more southern winter to their more northern summer quarters.

Spring Passage to Northern Europe. — The first appearance of the Thrush as a bird of passage takes place during the latter half of March,[1] when the birds which have wintered in South-Western Europe, and are en route for breeding quarters to the north of our isles, arrive on the south coast of England in company with Blackbirds, Fieldfares, Redwings, Wheatears, Warblers, Skylarks, Starlings, and occasionally Wood-cocks.

The passage continues throughout April, and down to mid-May, the voyagers, in company with Mistle-Thrushes, Ring-Ouzels, Redstarts, Blackcaps, Sedge-Warblers, and Corn-Crakes, in addition to the species already mentioned, pass northwards, chiefly along our eastern seaboard, and are joined, while en route, by many of our British emigrant Thrushes.

Such is the history of the Song-Thrush as a British migratory bird, when the tangled skein of its various movements has been unravelled and reduced to order. It is one that is only excelled in its complexity by a few species of British migrants, such as the Starling and the Skylark.

---

[1] From 19th to 26th March 1898, the Rev. O. Pickard-Cambridge, F.R.S., records an increasing number of thrushes around his rectory at Wareham on the coast of Dorset. On the 25th the land was fairly covered with them, and there must have been 200 or more in one field. On the 26th there were even more. On the 27th there were fewer, and by the evening of the 28th all had departed.—*Zool.*, 1898, p. 264.

CHAPTER XIII

In the British Islands the Skylark is not only one of the best known species, but also one which can be almost always met with, so that comparatively few people suspect the extent to which it is migratory, and fewer still are aware that the complexity of its migrations presents problems more difficult to solve than those of any other British bird, the Starling alone excepted ; yet this is undoubtedly the case.

As a migrant, no species makes so great a show in the returns from the light-stations, and the account which follows is based upon upwards of *five thousand* individual records. Yet within the British area the Skylark is for the most part resident as a species, though shifting its quarters when affected by frost or snow, as is obvious to almost any observer. The degree to which our native Skylarks are migratory depends on the varying conditions of climate and food. In the lowlands of Great Britain, especially in the south-west of England, and throughout Ireland generally, the migratory habit is less exercised, presumably because it is less necessary there than elsewhere. On the other hand, there are considerable tracts which, from their elevated, exposed, or northerly situation, are not suited for winter residence, and to these the Skylark is merely a summer visitor, as it is to nearly the whole of Northern and a great part of Central Europe, departing after the breeding season to its accustomed winter quarters. During its journeyings to the south and west in the fall of the year, and again on its return in spring, the Skylark appears in vast numbers on our coasts as a bird of passage, while, owing to their intermediate geographical position and their milder climate, the British Islands are much resorted to by the Continental Skylark as a winter visitant, one which largely replaces the native birds which have in the autumn fled our country.

The various migrations of the species may be conveniently separated and arranged as follows, beginning with the autumnal movements; and when it is considered that several of these movements are often simultaneously in progress, some idea of their complexity and the extreme difficulty of their interpretation may be realised :—

1. Local movements at the end of summer to winter retreats in the British Isles.

2. Autumn emigration of summer visitors (the members of the British migratory race), with their offspring—*i.e.*, home-breeding and home-bred birds.

3. Autumn immigration of winter visitors to England from Central Europe.

4. Autumn immigration of winter visitors to the British Isles from Northern Europe.

5. Autumn passage from Central to Southern Europe along the south-east and south coasts of England.

6. Autumn passage from Northern to Southern Europe along the British coasts.

7. Winter emigration from, and local migration within, the British Islands, due to severe weather conditions.

8. Spring immigration of summer visitors, and return of winter emigrants.

9. Spring emigration to Central Europe from south-eastern England.

10. Spring emigration to Northern Europe from the British Isles.

11. Spring passage from Southern to Central and Northern Europe along the British coasts.

12. Local movements in spring between British winter quarters and British summer (nesting) haunts.

But even this is not all, for the movements which take place between Great Britain and Ireland have also to be considered.

Autumn Emigration of Home-bred Birds.—Towards the close of the nesting season an increased number of Skylarks is observable in the lowlands, particularly near the coast; a fact due, no doubt, to migration from the higher grounds, to which the species is only a summer visitor. So early as July in some years there are a few records from the light-stations showing that departure has already commenced, but these early flittings must be regarded as exceptional.[1] During August there are usually a few signs of emigration, and towards the end of that month there is evidence that it has fully set in.

---

[1] A remarkable instance of this kind occurred on the night of 25th July 1881, when a great number of Skylarks appeared at the Leman and Ower lightship, off the Norfolk coast, and *sixty* were killed by striking the lantern, and at the same time *fifty* were killed at the Dudgeon, a neighbouring lightship. The weather was wet, changeable, and cold for the time of year. At the Eddystone, on the night of 31st July and through the earliest hours of 1st August 1902, Skylarks, Starlings, Wheatears, Sedge-Warblers, Curlews, and other species undetermined were crossing the Channel in some numbers. The wings of the species named, except those of the Curlew, were sent to me for identification, as they had been killed at the lantern of the lighthouse.

These late August movements include departures from the Hebrides and other western isles, as witnessed by birds observed at or killed against the lanterns of Skerryvore and Dhu Hearteach ; but there is no appearance of any emigration from Ireland in this month, which is a significant fact.

Throughout September the emigration is much more evident on both eastern and western coasts, the Hebrides contributing freely to the latter. A marked migration is also recorded from Shetland,[1] where the species is chiefly a summer visitant. In Ireland, too, there is evidence from the south-eastern stations that the exodus has begun. Towards the end of the month the movement is more marked, especially in unsettled weather, when Skylarks are recorded as emigrating by night in company with Thrushes, Blackbirds, Ring - Ouzels, Wheatears, Chiffchaffs, Whitethroats, Wagtails, Meadow - Pipits, Turtle - Doves, and other birds. As the season advances, emigration is naturally quickened, but it is impossible to fix the exact period at which the departures of our native Skylarks cease. After September these migrations become merged with the passage movements of the Continental Larks, also on their way southwards to winter retreats beyond our area. Our British Skylarks doubtless continue to quit our islands during October, along with the foreign birds. In some years a foretaste of cold, in others periods of exceptionally unsettled weather, cause pronounced "rushes" southward.

---

[1] The emigrations from Shetland, where the nesting season is late, commence during the third week of September, and they afterwards blend with the passage movements of the Larks proceeding southwards from Northern Europe.

During the autumn, Skylarks gradually draw towards the coast, on reaching which they pass southwards in straggling parties. On some days a succession of bands may be seen following each other throughout the whole day, and in September and October, if the weather be fine, with light winds, such congeries may be observed for days together without a break. This coasting movement is chiefly, if not entirely, performed by day; but it is otherwise when a considerable expanse of sea is to be crossed, as from Shetland, the Hebrides, or Ireland, for then the migration, as a rule, is undertaken by night. The journey is continued along both coasts of Great Britain until the southern and particularly the southwestern counties are reached, many of the east-coast migrants passing along the south coast westward. Probably only a portion of the Skylarks, which move during the early autumn, quit our shores, many, no doubt, tarrying on the south or south-western coast. Others, however, certainly depart for the Continent, crossing the Channel at many points, chiefly at night, in company with birds of many other species; but I myself, in passing between Newhaven and Dieppe in September, have observed small parties of Skylarks in mid-channel making for the French coast during the daytime. And at the Eddystone I saw parties proceeding towards France soon after sunrise. At Fair Isle, also in September, I have known them to depart in flocks during the early hours of daylight, flying towards the northern islands of the Orkney group.

Autumn Immigration from Central Europe.—This movement is the most interesting and remarkable performance of the Skylark, or perhaps of any British species, as it affords a striking instance of the phenomenon of birds proceeding westward and northward from their breeding grounds to reach their winter quarters, and this in vast numbers for several successive weeks, with scarcely a break, except those imposed by unfavourable weather. In some seasons this immigration—along what may be called especially the Skylarks' route, since they greatly outnumber the birds of any other species using it — sets in as early as the middle of September, but more commonly about the fourth week of that month. It is in October, however, that this stream of immigration becomes phenomenal. Some idea of the magnitude of this influx may be gathered from this table, showing the number of days during October on which it was observed in each of the years :—

1880. 22 days	1883. 9 days	1886. 23 days
1881. 12 „	1884. 19 „	1887.[1] 26 „
1882. 14 „	1885. 21 „	1903. (17th Sept. to 18th Oct.) 17 days

This stream has the coast of Essex and the mouth of the Thames for its centre, with its right wing extending to the Humber, or even beyond ; while the left sweeps south-west towards the Kentish littoral, and then along the south coast to Devon and Cornwall, some of the migrants crossing the Channel at various points to the French shores, while others continue westward and northward to Ireland, and appear on the coast of co. Wexford. The winter visitants to England, which are very many, among these immigrants, pass inland by several routes (numbers by way of the Thames and Humber estuaries), and disperse themselves over the eastern,

[1] Many recorded on 9th, 20th, 21st, 23rd, and 27th October ; vast numbers on 16th to 18th, again on 22nd, 25th, and 26th October.

southern, and midland counties.[1]   After October this
immigration falls off.   The November movements vary
according to the weather, but are never of great moment
after the first few days of the month, when in most
years they practically cease.

It is characteristic of this immigration that the
passage across the North Sea is chiefly witnessed during
the daytime, usually from dawn to noon, but not
infrequently onwards till 4 P.M., and that the birds concerned in it are actually crossing the line of flight taken
by the home-bred birds which are then emigrating—a
very remarkable occurrence, but not in October very
uncommon.   Other species which cross the North Sea
at the same time as the Skylarks are Starlings, Meadow-
Pipits, Tree - Sparrows, Redstarts, Willow - Warblers,
Chaffinches, and Rooks.

My investigations at the Kentish Knock light-
vessel in the autumn of 1903 lead me to believe that
Skylarks in numbers also reach our shores along this
route by night passages.   My reasons for this conviction
have already been stated when treating of the migrations of the Song-Thrush (p. 215).   During my sojourn
in the lightship, many Skylarks appeared at the lantern,
along with Wheatears, Redstarts, Spotted Flycatchers,
Tree-Pipits, Whitethroats, Willow-Warblers, Blackbirds,
Pied Flycatchers, Goldcrests, Chaffinches, Jackdaws,
Rooks, Mistle-Thrushes, Song-Thrushes, and Starlings.[2]
In addition to these movements from east to west,
Skylarks are observed at the North Goodwin lightship,

[1] It is a most remarkable fact that many of these Skylarks from Central
Europe, which winter in England, pass the cold season in latitudes *north* of
their summer homes.

[2] For the author's personal observations on these movements, see
Chapter XVIII., Vol. II.

off the east coast of Kent, flying from the south-east
to north-west — *i.e.*, from the northernmost part of
France across the Straits of Dover to Kent.

Autumn Immigration from Northern Europe.—
Great numbers of Skylarks, which summer in Scandinavia,[1] seek our shores in autumn for the purpose of
passing the winter with us. The date of the arrival of
these northern birds is remarkably constant—namely, in
the first week of October—when the birds appear in
Shetland, Orkney, on the east coast of Scotland and
north-east coast of England, during the night or early
in the morning, in company with Thrushes, Redwings,
Blackbirds, Ring-Ouzels, Goldcrests, Chaffinches, Bramblings, Redbreasts, Starlings, and other species. These
arrivals continue at intervals during the month, and
the Skylark participates largely in those remarkable
movements which characterize October's latter days.
These vast outpourings seem to exhaust the emigration from Northern Europe, for it is doubtful if any
considerable arrival takes place in November even in
its earliest days. Thus the autumnal immigration
from the north, vast as it is, is compressed, as it were,
into the period of some four weeks. The majority
of the northern Skylarks disperse themselves over our
islands (including the Hebrides, the Shetlands and
Orkneys), and replace those home-bred birds which
have quitted their summer haunts. A great many
seek Ireland by direct passage from south-west Scotland ; others, perhaps, by way of the Isle of Man,
or from the Welsh coast to the shores of co. Dublin

---

[1] Professor Collett says (*Oversigt af Christiania Omegns ornithologiske
Fauna*, p. 128) that *Alauda arvensis* is seldom seen in the Christiania
district after the middle of October.

and co. Wicklow. Not unfrequently in October and early November flocks are observed passing Scilly in a north-westerly direction, *i.e.*, towards the Irish coast.

Autumn Passage from Central and Northern Europe to Southern Europe. — These movements are much involved with the immigratory movements from the East and North, and to a lesser degree, with the British emigratory movements. The Skylarks participating in these great passage flights from the North arrive in October and early November at our northern islands and on our north-eastern coast, in company with those which winter with us, whose migrations have just been considered. After a short rest, these travellers proceed along the coasts, chiefly by night, southward and westward to cross the Channel at various points. Though they are mainly observed on our eastern and southern seaboards, yet a considerable number pass along the west coast, visiting annually such outlying islets as Sule Skerry, St Kilda, the Flannan and the Monach Isles, and traversing the Outer Hebrides from end to end. Others reach Ireland, and continue their southerly journey along its eastern and western shores. The passage movements from Central Europe by an east-to-west flight across the North Sea to the south-east of England need no further notice now, since they have been treated of already under Autumn Immigration from Central Europe. Probably a number of these transient visitors tarry for a time with us ere they proceed southwards, for Skylarks departing from southern England are observed at the Eddystone down to the middle of November.

General Remarks on Autumn Emigration and Immigration.—Having treated of the autumn movements, both of emigration and immigration, it may be desirable before proceeding further to consider their effects on the Skylark population of Britain, and its position at the end of that season. Though vast numbers of home-bred birds have at that time quitted our shores, their departure has not materially affected the great abundance of the species, partly owing to the fact that the Skylark is double-brooded,[1] and hence its annual increase is enormous, while prodigious numbers have poured into England from Central Europe during part of September and throughout October, to say nothing of the immense number of immigrants from North-Western Europe to the British Isles generally, which have arrived during the latter month. The result is, that from November to the setting in of cold weather, the Skylark population of the British Isles is at its maximum, and vastly in excess of what it is at any other period of the year.

During these great and varied autumn movements, immense numbers of Skylarks perish at the lanterns of the numerous light-stations both on and off our shores. Indeed, no other bird is so great a martyr to the allurements of the beacon light. My experiences of some of these terrible immolations are related on pp. 291 and 344.

Winter Emigration from, and Partial Migration within, the British Islands.—These movements depend wholly on the state of the weather, and vary in degree according to its severity. The Skylark, obtaining the whole of its food on the ground, is at once driven to

---

[1] In many parts of England most pairs of skylarks have three nests in the year.

change its quarters when that is covered with snow, and only somewhat less quickly when it is merely frost-bound without snow. Should the late autumn and winter be uniformly mild, the Skylarks sojourning with us remain practically stationary. Few if any winters are, however, entirely free from snow or frost, and with the first outbreak of cold the birds must remove themselves from its untoward influence. Sometimes suitable lodging may be found not far off, and then the movement is but local or partial in character. When this occurs, and the stress is but short, the birds soon return to their former haunts; but if the adverse conditions continue and become general, the movement also becomes widespread and more or less universal. This effect is especially produced by great snowstorms, when the number of fugitives is so vast that people wonder where such prodigious multitudes can come from, as they throng towards the coast, and sweep along the seaboard and its neighbourhood to reach the milder south-west coast of England—Devon, Cornwall, and the Scilly Isles—though many undoubtedly cross the Channel for the Continent, and others proceed to Ireland. On the other hand, a few—and these are, perhaps, of our native stock—attempt to brave the unfavourable conditions, partly by resorting to unwonted places of shelter, especially the sea-shore, but many, if not most of these, succumb to famine. In Ireland, too, there are many winter movements, due to the pressure of climatic conditions, and Cork and Kerry are especially resorted to during hard weather; but winter emigration must be regarded as exceptional in Ireland, for one portion or another of its shores generally affords an asylum in the severest seasons, though many birds perish, even in its most favourable areas, during an abnormally protracted winter. It has already been stated that Ireland ordinarily receives numbers of Skylarks in autumn, and as it is again sought by multitudes of refugees from the snows and frosts of Great Britain, it follows that the Skylark population of Ireland is at its maximum at a period when that of Great Britain is at its lowest.

When the winters are exceptionally severe on the Continent, there is a renewal of the movements of Skylarks (together with Starlings and Lapwings) across the southern waters of the North Sea to the south-east coast of England.

During these cold-weather movements, many of the emigrants perish at the lanterns of the light-stations. Thus, on 2nd December 1882, the Bell Rock light-house was visited by what is described as being the greatest multitude of Skylarks ever known. It was impossible to estimate the number, but they were "striking hard for a couple of hours like a shower of hail."

If the statement that the winter emigration depends wholly on the state of the weather needs any confirmation, it may be furnished by the fact that in the mild seasons 1881-82 and 1885-86 very little was recorded. There are, however, usually spasmodic and partial movements in November; but it is not until cold weather sets in that any general exodus takes place. If there has been much snow in December, as in 1879 and 1882, there is little or no movement later in the season, because the birds have already departed. On the other hand, after the uneventful December of 1880, there were pronounced emigrations in January 1881, when a cold period set in.

In February there are, as a rule, movements more or less local, and due to snow, and in that month of 1886, which was cold and snowy, movement followed movement throughout its course. The March migrations are not of much account, since they are local in their nature, but in unusually inclement seasons, like 1883 and 1887, there were "rushes" to the coast as late as the 20th of that month.[1] For further instances of the effects of severe seasons on this and other species, see the chapter which is devoted to Winter Movements.

Spring Immigration of Summer Visitants and Return of Winter Emigrants.—The return of the Skylarks which have left us during the autumn and winter is observed on the southern coasts of both Great Britain and Ireland early in the year, their arrival beginning, as a rule, during the latter half of February, and occasionally as early as the second week (in 1886 on the 11th), the immigration continuing throughout March. The precise time seems to be influenced by the character of the season. If the early spring be mild and genial, they begin to return early; but if the contrary, their appearance is delayed. On arrival on the south coast of England many pass northward along the east and west coasts, the latter being the route chiefly followed by the earlier immigrants. The return to Ireland corresponds closely with the arrival in southern England, the earliest observation for the period 1882-87 being on 10th February 1886, and from that time the movements occur at intervals. The other species of birds which

---

[1] At the Nash lighthouse, on the Glamorgan coast, on 15th March 1887, Skylarks, Starlings, Snipes, Woodcocks, Lapwings, Golden Plovers, Wild Ducks, and others were seen flying before heavy snow from 8.30 A.M. to 3 P.M.; two or three hundred being seen at a time.

reappear along with the Skylarks are mostly those which have before been mentioned in association with them — Thrushes, Blackbirds, Mistle-Thrushes, Red-breasts, Meadow-Pipits, Starlings, Lapwings, and so forth.   During March the movements of the immigrants become merged into those of the birds of passage strictly so called ; and arrivals on the south coast usually cease in the first week of April.   In Ireland, during the first days of the month (April) and occasionally to its third week, Skylarks continue to arrive in company with Wheatears and other early summer birds.   The return movement to the Hebrides corresponds with that to the mainland, but, as in Ireland, the immi-gration is prolonged into April.   In Shetland the spring arrival of the native birds begins in the early days of March, in some seasons in the last days of February.

The immigrants reach the south coast of England, sometimes in vast numbers, during the earliest hours of the morning ; but in the south-east of Ireland, the chief point of arrival in that country, they are usually observed later in the day—in the Hebrides mainly at night.

Spring Emigration from the British Isles to Central Europe.—The return (west to east) movement from south-eastern England across the North Sea comes very little under observation compared with the in-flowing streams of the preceding autumn, and that this should be so is easily explained.   In the first place, the numbers of travellers, owing to the waste of winter, have been much thinned ; and secondly, like most other important emigratory movements, this one takes place chiefly at night, and so for the most part escapes notice, for it is reasonable to suppose that the first hour of flight carries the birds beyond the limit of observation at the light-stations off our eastern coast. Some return emigration is nevertheless observed by day from the lightships, the direction of the birds being eastward from the mouth of the Thames, and south-eastward from the more northerly stations. There are also enough observations to show that the movement begins in February (in the mild season of 1882 on the 6th, but usually not till the middle of the month), and is continued until the end of March, the 28th being the latest day recorded. As with the reverse movement in autumn, this is chiefly noticed from the lightships between the Thames and the Humber. The other species of birds accompanying the emigrant Skylarks are Starlings, Rooks, Grey Crows, and Lapwings.

Spring Migration from the British Isles to Northern Europe.—In mild seasons, during the third week of February there are indications at our north-eastern stations that the Skylarks which have wintered with us are beginning to depart for their northern homes (including the Orkney and Shetland Islands), and throughout March, especially after the middle of the month, and during the first week of April, there is usually much evidence to the same effect, the concomitant species being Blackbirds, Goldcrests, Starlings, Woodcocks, and "Wild Geese."

The spring emigration from Ireland deserves separate consideration. Beginning about the middle of February, it becomes more pronounced in March, and ceases with the close of that month. The birds return by the routes taken in autumn and winter, chief of which is that between the south-eastern counties, with Wexford as a centre, and the southern province of Wales and shores of the Bristol Channel; while during March there are return flights across the Irish Sea to north Wales and south-western Scotland. Generally, the birds set out after dark, but Skylarks are occasionally recorded as migrating during the day, those from the southern portion of Ireland making for the south-east, while those from the Wicklow coast proceed due east. The night movements are often performed in company with Thrushes, Blackbirds, and Starlings. The winter visitants to the Hebrides leave for the mainland of Scotland about the same time, and call for no special remark.

Spring Passage from Southern to Northern and Central Europe along the British Coasts.—These movements take place during March and early April, but there are Eddystone records dating as late as the third week of the latter month, when the last of the Skylarks appear at that station in company with Wheatears, Ring - Ouzels, Willow - Warblers, Fieldfares, and Redwings. It is probable, however, that the bulk of the Skylarks arriving at this time on the southern coast of England are en route for North-Western Europe. After reaching our shores, they move northward along the coasts, and finally quit the country in company with those which have been wintering in Great Britain and Ireland, as well as with emigrants and transient visitors of other species.

Local Spring Movements of Native Birds.—Those home-bred Skylarks which left certain districts in the autumn in search of more suitable retreats in our islands in which to pass the winter, return to their nesting haunts as early as February in genial seasons, but March is the main month for such return movements.   Such immigrants are, however, subject to eviction on a recurrence of winter conditions, but return at once when the storm is over.

CHAPTER XIV

THE MIGRATIONS OF THE LAPWING,
VANELLUS VANELLUS

THE migrations of the Lapwing in the British Islands are especially interesting, and possess features which are not shared by any of the species previously treated of.

When we come to investigate the various movements performed by this well-known bird, it is surprising to find how largely some of them escape notice. This may to some extent be accounted for by the fact that the Lapwing is partially nocturnal in habit, and hence less prone than most species to approach the light-stations or otherwise come under notice, when performing some of its most important movements; but this does not, I think, afford complete explanation, for other *Limicolæ*, of even more pronounced nocturnal proclivities, such as the Woodcock, do not pass unobserved to a like degree.

In addition to being a resident during the major portion of the year in extensive areas of our islands, the Lapwing is chiefly, indeed almost entirely, a summer visitor to Shetland, Orkney, the Hebrides, the more elevated districts throughout the mainland of Great

Britain and Ireland, and to other inland areas, especially in the north—facts which result in much migration taking place within the British Islands, and of emigration to countries south of us.

As winter visitors and birds of passage, some numbers arrive on our shores in the autumn from Scandinavia and from Western Central Europe.

The great majority of our home-bred birds, and perhaps also of the continental immigrants, pass the cold season in Great Britain and Ireland ; but extensive winter movements are performed under the pressure of severe climatic conditions, which affect the food supply. At that season many emigrate southwards and westwards, and others cross the Channel to the shores of France.   A number of minor or local movements, due to varying weather influences, are also performed in both autumn and winter in all parts of our islands resorted to by this bird.   Indeed, almost every decided change in the weather results in some shifting of quarters at these seasons.

Our native Lapwings are widely distributed and extremely numerous, and there can be little doubt that they form the great majority of the individuals which participate in the many and various migratory movements undertaken by this species at different seasons.

Except the winter movements, when forced and sometimes general retreats have to be undertaken, the migrations of the Lapwing are very gradually performed and cover an extended period in each season ; but no general flights simultaneously performed over any large section of our coasts have been recorded.

Summer and Autumn Movements within the British Islands.—At the close of the nesting season, Lapwings old and young gather together and form flocks. As early as mid-June, or during July, small parties, even flocks, sometimes leave their nesting grounds to appear in the vicinity of the coast, and occasionally a few are recorded as visiting the islands, or as occurring at the rock-stations and lightships off the east and west coasts of Britain. Such local movements are not without interest; but as a rule it can scarcely be claimed for them that they possess any direct bearing upon the ordinary migrations of the bird. During the second week in July, Lapwings from Shetland appear in small parties at Fair Isle, and movements southward have been recorded;[1] but these must be regarded as exceptional, and in some cases are possibly due to the disturbing influence of local meteorological conditions.

In August emigration from Shetland continues, and many leave during the month. The records of Lapwings at the coast stations and at the off-lying lighthouses and lightships[2] are numerous, though as yet irregular and uncertain, but they indicate that movements or wanderings are in progress.

In September the migration southwards sets in in earnest. Early in the month, those which have summered in Shetland, Orkney, and the Hebrides, continue or

---

[1] The chief of these was observed at the Leman and Ower lightship (twenty-five miles north-east of Cromer) on 30th July 1887, when, during unsettled weather, a great flock passed southwards at 3 P.M. On 2nd July 1901, after a strong north-east wind and dirty weather, Mr T. Southwell saw on board a Lowestoft trawler two which had been captured out of a large flock about forty miles north-east of that port.

[2] At the Seven Stones lightship (seven miles off the Land's End), on 26th August 1880, about fifty lapwings were observed flying south-west in the direction of the Scilly Isles at 10 A.M.

begin to emigrate, and by the middle of the month, or before, all save a few stragglers have departed from the northern group. Many, too, quit the higher ground on the mainland, especially in Scotland. These decided emigrations result in movements southwards or towards the shore, which are chiefly in evidence on both the coasts of Scotland and those of the north of England.[1]

During October the autumnal movements reach their maximum, while a number of foreigners arrive whose movements will be treated of immediately. The higher breeding grounds are then entirely deserted for the cold season, and much emigration is also in progress from the northern and inland districts and from the Hebrides to accustomed winter quarters, such as lowlands in the vicinity of the coast—especially near estuaries—and the southern counties generally, including the Scilly Isles, which are annually resorted to from October onwards. Many, however, remain during mild winters in suitable haunts in northern Scotland, as in the neighbourhood of the Beauly and Moray Firths, where the climatic conditions are exceptionally favourable. The British October migrants are observed on all sections of the coast, and the movements southwards are no doubt augmented by the presence of immigrants from Northern Europe.

The movements during the first half of November are a continuation of those of October. By about the middle of the month, the birds have usually settled down for the winter, or until they are compelled to move by

---

[1] At the Inner Farne Island, on 26th September 1882, "thousands" are recorded as having appeared, along with a few Golden Plovers and many Curlews.

the pressure of adverse weather and its effect on their food supplies. A small number remain in the Hebrides until the first onset of frost, the species being resident there in mild seasons.

Although a few records, both Scottish and Irish, point to some immigration into Ireland from northern Britain in September, it is not until October and November that such movements are regularly observed. The main lines of intermigration lie between the Mull of Cantyre and the Solway and the coasts of Antrim and Down; while birds quitting or traversing the Hebrides reach the shores of Donegal, some of them by way of Tory Island. During October and November, too, there is some evidence of the arrival of Lapwings in Ireland from the south-east by a passage, chiefly observed during the day-time, across St George's Channel to the Wexford and adjacent coasts. Certain of the later November Irish immigrations are associated with the setting in of more or less severe weather in Scotland.

Autumn Emigration from Britain.—A number of our native Lapwings quit the southern shores of England late in September, departing at night in company with other summer visitors, such as Ring-Ouzels, Wheatears, Whitethroats, Grasshopper-Warblers, Sedge-Warblers, Pied and Spotted Flycatchers, Pied and Yellow Wagtails, Turtle-Doves, etc., etc.[1] These departure movements continue during October, when the British Lapwings are joined by the autumnal visitors which have arrived from the Continent.

[1] A Lapwing, "ringed" as a nestling at Glenorchard, Stirlingshire, on 17th June 1909, was shot on 17th November about twenty miles west of Pau in south-west France. Another "ringed" British Lapwing was recovered in Portugal.

These cross Channel migrations are not much in evidence in the records, however, until November, when no doubt the approach of winter, with its low temperatures, constrains some of them to seek more genial climes. During this month they have been observed leaving our southern shores, especially at the Eddystone and the Isle of Wight stations, in considerable numbers at night, in company with Mistle - Thrushes, Song - Thrushes, Fieldfares, Redwings, Blackbirds, Starlings, Larks, Golden Plovers, and others

Autumn Immigration from North-Western Europe. —The autumn arrival of Lapwings on our shores from Scandinavia[1] sets in during the first week of October and lasts a little over a month. It is observed at stations extending from Shetland to the northern section of the east coast of England. No great arrivals, covering extensive portions of the coast-line, have been recorded, as in the case of other species, but only scattered instances of moderate numbers (a hundred or so) appearing at intervals. At the northern islands, where the birds arrive some time after the summer visitors have departed, their appearance is irregular; in some seasons they occur in fair numbers, while in others they are very scarce. In Shetland, after the early days of November, stragglers only are observed. It is doubtful if we derive any very great numbers from Northern Europe, as that portion of Norway from which the British Isles presumably receives immigrant Lapwings affords only

---

[1] The Lapwing is only a rare straggler to Iceland, but has been observed on passage in small numbers during both spring and autumn, in most years, at the Faroes (Andersen).

somewhat limited haunts for this bird as a summer visitor.

The northern immigrants arrive on our shores during the night and the earliest hours of the morning, and frequently appear simultaneously, if not in company, with Song-Thrushes, Fieldfares, Redwings, Blackbirds, Ring-Ouzels, Redbreasts, Goldcrests, Bramblings, Skylarks, Starlings, Snipes, Woodcocks, etc.

Many, perhaps the great majority, of these northern birds are bent on passing the winter in our islands, and proceed along the coastlines and move inland to reach suitable haunts. Whether they remain the entire season depends upon the nature of the winter—*i.e.*, whether it is sufficiently mild to allow them to do so.

Autumn Immigration from Western Central Europe. —During the last days of September, throughout October, and in early November[1] there are records of parties of Lapwings being observed at the lightships off the south-east coast of England. These migrants are proceeding in a westerly direction towards the coast of Essex and the mouth of the Thames as a centre, but with their right wing extending to the Norfolk, Lincolnshire, and even Yorkshire coasts, and their left to the east coast of Kent. These Lapwings from the opposite coast of the Continent, reach our shores by a direct passage across the southern waters of the North Sea. The arrivals take place during the daytime and at night ; but there are no general movements recorded, and the observations chronicled are few and scattered during any

---

[1] In 1885 at the Hasbro lightship, off the Norfolk coast, many were passing west and west-north-west on the nights of 22nd and 23rd November— perhaps a cold-weather movement from the Continent.

season, though the numbers recorded are occasionally considerable.

On arriving on the south-east coast of England, the majority of the birds pass inland at various points along the eastern and southern seaboard, and many pass up the Thames and Humber estuaries to reach the interior ; the object in all cases being to find congenial winter quarters.

Autumn Passage of Immigrants Southwards beyond the British Isles.—On reaching the British Isles from the north or east, those immigrants which do not intend to winter here pass on to the southern coast, whence they cross the Channel, bound further south.

During these passage movements, some of the birds from the north which travel by the west coast visit such remote stations as Sule Skerry and the Flannan and Monach Islands.

Winter Movements and Emigration. — The winter movements of the Lapwing consist of emigrations from Britain for more southern lands, and of partial or extensive migrations performed within our area. They are controlled by, and vary with, the climatic conditions of the season, and their extent is proportional to its severity. Should the late autumn and the winter prove mild, the Lapwings remain unmolested, so to speak, in their accustomed retreats. Sooner or later, however, cold weather of a more or less severe type, and of either local or general prevalence, sets in, and then the birds, owing to their inability to obtain food, are compelled to change their quarters for others free from its blighting influence : these havens may be near at hand or far removed, in accordance with the area adversely affected. Heavy snow and severe frost cause great movements south-

wards along the coasts and overland. Should such
conditions extend to the south of England, much
emigration for the shores of France [1] follows, great
numbers of Lapwings crossing the Channel both by
day and night. Occasionally during these periods of
exceptional severity many of these birds, along with
other species affected ("Thrushes," Larks, Starlings,
etc.), are observed moving westwards during the daytime
along the south coast of England and its vicinity en
route for Devon, Cornwall, and the Scilly Isles,[2] and
not a few then cross St George's Channel to Ireland,[3]
where milder conditions usually prevail. In the Sister
Isle the counties of Cork and Kerry are largely resorted
to by the Irish birds when in distress. There appear to
be no winter movements westward to the Hebrides, as
there are in the case of several other species affected by
severe weather on the mainland.

The time when the winter emigrations from Britain
may be enforced varies greatly. Thus, during the
season 1901-1902, the weather in the south of England
remained mild until February, when it became excep-

[1] On the 6th of December 1902, several hundreds arrived at the Eddystone at 7.15 P.M. and were flying in the rays from the lantern until 5.45 A.M.
On the following night they appeared again at 7.30 in still larger numbers,
and were striking the copper dome of the lantern so continuously that "it
was more like a maxim gun at work, and they were falling overboard by the
score. The thermometer stood at freezing-point, a very rare occurrence
here."

[2] On 13th February 1900, a great flock passed over these islands coming
from the north-west. The next day the islands were alive with an extraordinary assortment of Lapwings, Golden Plovers, Starlings, Song-Thrushes,
Mistle-Thrushes, Redwings, Fieldfares, and Blackbirds. They passed on,
but left many dead.

[3] Occasionally during severe winters numbers of Lapwings have been
observed passing westwards at stations off and on the east coast of Ireland,
which points to a passage of emigrants from north Wales.

tionally severe, and so continued for long.   At the
Eddystone and elsewhere on the south coast, no Lapwings had been observed crossing the Channel previously
during the winter, but on 2nd February, and again on
the 13th and 15th, great numbers passed southwards
both by day and by night.   Winter movements within
our isles have been recorded as late as mid-March; on
the 15th, in 1881, many Lapwings, along with Skylarks,
Starlings, Golden Plovers, Woodcocks, and Snipes, were
observed flying southwards before snow at the Nash
lighthouse, on the north shore of the Bristol Channel.

It is when retreating before these adverse conditions
that the movements of the Lapwing become pronounced
and widespread, and in this respect contrast markedly
with the other migrations of this species.   It is on such
occasions, too, that the bird chiefly approaches the
lanterns and is killed or captured—a fate which does not
commonly befall it.[1]   On the night of 17th December
1885, twenty-one were captured at the Eddystone out
of several hundreds which appeared at the light—the
record for the period 1880-1887.

Winter Immigration from Western Central Europe.
—During severe winters on the Continent, the east-to-
west passage of Lapwings across the southern waters of
the North Sea is renewed.   The immigrants, as in the
autumn, arrive on the south-east coast of England, and
pass westward in search of the milder areas within our
isles, or proceed further south beyond our shores.[2]

---

[1] In sixteen years eight only were obtained at the Irish light-stations;
and Mr Herluf Winge informs me that twenty-one were killed at the Danish
stations during a like period.

[2] At Great Yarmouth, on 22nd December 1894, hundreds of Lapwings
were observed "coming over" against a strong north-west gale, and many
were drowned.—*Zoologist*, 1900, p. 163.

Spring Movements from British Winter to British Summer Haunts.—Lapwings may be induced by the prevalence of mild weather to return in small parties to their breeding-grounds in England and the south of Scotland as early as the end of January and beginning of February, but are usually compelled to retreat again by the recurrence of severe climatic conditions. The usual period for their appearance in their summer haunts is about the last week in February and early in March; and should severe weather follow in the last-named month, many then perish in the more exposed areas. The Scilly Isles, where many winter, are quitted by the middle of February in ordinary seasons.

Immigrants occasionally appear in Orkney at the end of February; but this is exceptional, for the summer visitors to these islands do not usually arrive before March. The return to Shetland is timed with great regularity for the first week of March, and, as in Orkney, the arrivals are in progress throughout the month. The Hebrides are sought during late February and early March. The spring movement is a gradual one, as a rule, but on 7th March 1908 over six hundred arrived in a flock at Fair Isle, doubtless en route for the Shetland Islands.

Spring Immigration from Southern Europe.—The return movements of the Lapwings which departed from our islands in the autumn and winter are observed at the light-stations and elsewhere on the south coast of England. Like the movements of the autumn, they do not come much under notice. Fortunately, however, we possess some important records for the latter half of March and during the first half of April, when the Lapwings have been observed returning in company with Wheatears, Mistle-Thrushes, Fieldfares, Black-birds, Chiffchaffs, Willow-Warblers, Redstarts, Skylarks, Starlings, and other species. The chief of these cross-Channel return movements were witnessed during the earliest hours of the morning, and the Eddystone is one of the main stations for their observation.

Spring Return to Ireland. — Late in February, during March, and sometimes early in April, Lapwings are observed during the daytime arriving from the south-east and passing north-west at the light-stations off the Wexford coast. There is no special reason for regarding these as passage movements, and, taken together with the facts (1) that they are not observed proceeding along the eastern coastline northwards, and (2) that during winter Lapwings are recorded as passing southwards on the extreme southern sections of both coastlines, it is not improbable that some of these birds quit Ireland under the pressure of climatic conditions and return in the spring.

Spring Emigration and Passage Northward to Northern Europe.—The spring movement northward to Continental breeding-haunts is one of the best-observed phases in the ordinary seasonal migrations of the Lapwing. It is witnessed on both the east and west coasts and at the northern islands.

The spring emigratory movements from the mainland begin with the departures from their winter quarters of Lapwings returning to their nesting-haunts in the Orkneys and Shetlands. In some years many have been observed during March, but these early migrations are dependent upon the genial nature of the season. In March the local British migrations, if we may so term them (for the birds are returning to a remote part of the

British area), become merged with the migrations of birds departing from our islands for Northern Europe, and with the passage of birds along our coasts proceeding thither from winter quarters south of our area. These movements are regularly observed, and are much in evidence during April, after which stragglers are observed throughout May and even during the first week of June. "Rushes" are recorded, when Lapwings, in company with Fieldfares, Redwings, Ring-Ouzels, Blackbirds, Pipits, Redbreasts, Whinchats, Redstarts, Pied Flycatchers, and Ortolan Buntings are observed at the Shetlands en route for the north.

These movements take place during the night, and the resting migrants are observed during the daytime at various points on the coast and the off-lying islands—the Farnes, the Isle of May, and the Pentland Skerries on the east, and the Isle of Man, Dhu Heartach, Skerryvore, Sule Skerry, the Flannans, and occasionally even the remote St Kilda on the west—and at the various Isles of the Orkney and Shetland groups. Our departing winter guests, which form the bulk of these birds, proceed overland in various directions to reach the coast of the mainland, whence they proceed northwards; many of those travelling along the east coast cross the North Sea ere the Orkneys and Shetlands are reached, though considerable numbers visit these islands regularly on passage to their Scandinavian summer quarters.

The Lapwings which have wintered in Ireland begin to move northward in mild seasons, about the middle of February; but the chief emigrations take place during March, after which month they fall off, though some have been observed in the end of April. The latter, however, are exceptional occurrences. The emigrants mainly depart from the north-east coast and proceed in various directions towards Scotland, and are chiefly recorded for the daytime. There is also an emigration to the south-east from the Wexford and adjacent coasts late in February; but this, like most emigratory movements, largely escapes notice, and our data regarding it are only slight.

Spring Emigration to Central Europe. — This movement has not come much under observation, a circumstance which is not surprising. Such departures from our shores are embarked upon under favourable weather conditions and after nightfall, and the birds are not likely to come under notice at the fleet of lightships off the south-east coast of England—the only observatories—immediately after departure. There are, however, a few important records which clearly indicate that Lapwings pass eastwards during the latter half of March, and at night-time.

CHAPTER XV

THE Starling is a summer visitor to Northern and much of Central Europe, and a winter visitor to Southern Europe and Northern Africa. In the British Isles it is a resident, a local migrant, a summer visitor, a winter visitor, and a bird of passage.

The migrations of the Starling observed in Great Britain and Ireland are of a singularly varied nature, being performed with great frequency and at all seasons. These remarkable characteristics in the movements of a well-known and familiar bird are due to a number of causes—among others, to its gregarious and capricious nature, the varying degree of its migratory instincts in different parts of the British area, its dependence upon supplies of food which change not only with the season but from year to year, and to the fact that it is largely double-brooded—peculiarities which result in innumerable movements, many of them of a partial or wholly irregular nature.

In addition to these, there are the regular migrations performed by the Starling—(1) as a migratory species within the British area; (2) as a summer visitor from the south; (3) as a winter visitor to our isles from Northern and Central Europe; (4) as a bird of double passage, traversing our shores when en route between Continental summer and winter quarters ; and, finally, (5) there are winter movements—partial migrations within the British area and emigration to the Continent—dependent upon and varying with the severity of the season.

The data amassed relating to these numerous irregular and regular movements are extraordinarily voluminous, and their study has presented problems for solution of an exceptionally complex nature—more so than those appertaining to any other British bird.

As a resident species the Starling is widely distributed over our islands, its range extending from the Shetland and other northern isles[1] southward to the English Channel. It is also a permanent resident at St Kilda. In many of the northern and of the more elevated portions of the mainland of Britain the bird is migratory, being entirely or partially absent during the autumn and winter months.[2]

This variability in the migratory habit is also manifest in many districts of England. It may in most cases depend upon the distribution of food-supplies ; but this does not explain all, for there are counties in south-western England, such as Cornwall and Devon, in which the Starling has only recently become a breeding species, and is still chiefly a winter visitor.

In Ireland the peculiarities in seasonal distribution of native Starlings are very similar, and the species is mainly a winter visitor to the south and west. An

[1] At Fair Isle there appears to be no diminution in the numbers of native birds in winter ; but in north Ronaldshay, one of the outermost and most exposed of the Orkneys, only a few remain for the winter.

[2] At Halmyre, a moderately elevated district in Peeblesshire, about 75 per cent. leave (Laidlaw). At Pitlochry, in Perthshire, which is flanked by high ground, all depart (Macpherson).

interesting and important fact is that in Ireland winter visitors from Great Britain and the Continent far out-number the Irish birds.

Summer and Autumn Movements of British Star-lings.—These take the form of (1) local migrations within the British area, and (2) of emigrations of native birds to winter quarters beyond our shores.

1. *Local Migrations.* — These begin in the early summer; indeed, as soon as the young, especially those of the broods first cast off, are able to shift for them-selves. Sometimes as early as the first week in June parties composed mainly of youngsters commence their wanderings; but it is usually about the middle of the month that such flocks are commonly observed. Even thus early the maritime districts, the light-stations, and islands lying off the coast are sometimes visited. Later in the summer both old and young gather together and form-large flocks. Movements of a more definite nature are then undertaken, at first probably in search of fresh feeding-grounds, and finally directed towards winter homes.

The coast and its vicinity is largely visited, especially the southern and western seaboards; and when summer is past the Hebrides and other islands (including Scilly) and Ireland are also sought for the winter. These movements commence in some seasons as early as the end of July,[1] and are in progress throughout the autumn. Ireland receives considerable numbers of immigrants from England, Scotland, and Wales, towards the end of August and onwards.

---

[1] At the Tuskar Rock, off the south-east coast of Ireland, on 27th July 1894, several Starlings were observed proceeding in a north-westerly direction —*i.e.*, making for the Wexford coast.

Later in autumn these movements merge into those of the Continental hosts also seeking winter retreats in various parts of our islands.

2. *Summer and Autumn Emigration.*—Not only are winter quarters sought by our native Starlings within the British area, but many travel much further to find retreats in South-Western Europe.[1] Thus a number of our British-breeding Starlings are summer visitors to our islands.

Late in July, during August, and up to the middle of September (before the Continental birds appear on our shores) emigrant Starlings depart from the south coast of England, and are observed crossing the Channel towards France, sometimes in company with Wheatears, Sedge-Warblers, Song-Thrushes, Meadow-Pipits, Sky-larks, Curlews, and other species. These movements of departure are performed during the night or the earliest hours of the morning, and hence for the most part escape notice; but I have received during the past few years much valuable information regarding them from the Eddystone lighthouse, the situation of which is singularly favourable for the making of such observations. Nearly all the Starlings (and other species) which meet with an untimely end at the lanterns at this season are birds of the year—a circumstance, however, to which no great significance should be attached; for we must remember that the majority of the emigrants are young, indeed only a few weeks old, and it seems natural that they should fall easier victims to the attractions of the lanterns than older travellers with more experience. The later British emigrants doubtless depart with the

[1] Marked native British Starlings have been recovered in winter in France.

Continental birds which have traversed our shores on their way southwards.

Some of these native emigrants are probably of Irish origin, but their departure is likewise difficult to detect. There are, however, nocturnal movements (and emigratory movements are eminently performed by night) of Starlings and other birds during the latter part of July and in August, which seem to indicate that this species quits Ireland in the late summer and early autumn for more southern winter quarters.

It is possible that some Starlings may cross the English Channel in the daytime. There is, however, but one record of such a movement in the returns;[1] and during a five-weeks' residence at the Eddystone in September and October 1901, I never saw any diurnal migration on the part of this species, though many thousands crossed in the night and earliest hours of the morning.

Autumn Immigration from Central Europe.[2]—The first Starlings to arrive from the Continent on our coasts in the autumn come from the east, and are doubtless emigrant summer visitors from Western Central Europe. These visitors cross the southern waters of the North Sea by a more or less direct east-to-west passage, and appear on the coast of England from the Humber southwards to the Channel.

These immigrations set in with great regularity during the last week of September,[3] reach their maximum

[1] At the Varne lightship (Straits of Dover), on 18th September 1887, twenty passed from north to south-south-east at 7 A.M.

[2] For further information regarding the migrations of the Starling along this route, as observed at the Kentish Knock lightship, see Chapter XVIII., Vol. II.

[3] The earliest date chronicled is 21st September 1880, but the initial date for other years follows closely thereon. I observed the first at the Kentish Knock lightship on the 24th in 1903.

volume in the last three weeks of October, and usually
cease with the early days of November; but in some
seasons there are arrivals until the middle of the month.[1]

As an illustration of the magnitude of these inpourings,
it may be stated that they have been recorded for as
many as twenty-one days during October, and that the
chief "rushes" often cover several successive days, and
affect the eastern coastline from the Humber southwards.
The passage is chiefly performed during the daytime,
and not unfrequently lasts from early morning until dusk,
sometimes under most trying weather conditions;[2] but
there are records which doubtless refer to night passages.

As in other immigrations along this route, the direction of flight varies, being from direct east to west at
its centre about the mouth of the Thames, to the south-
west off the coast of Kent, to the north-west on the
Norfolk coast, and to the north-north-west at the mouth
of the Humber.

The species which have been observed migrating
from east to west on the same dates as the Starling are
Rooks, Jackdaws, Skylarks, Tree-Sparrows, Chaffinches,
Meadow-Pipits, and Lapwings.

Many of these immigrant Starlings from Central
Europe winter in various parts of England:[3] many, too,
pass along our southern shores; some to cross the
English Channel at various points on their way to
retreats in South-Western Europe, while others proceed
to Ireland, where they arrive on the coast of Wexford

---

[1] Latest at the Corton lightship on 17th November 1880.

[2] At the Leman and Ower lightship, on 24th October 1884, a flight
estimated at five thousand passed landwards at 5 P.M., and of these fifty
struck the lantern and were killed.   (See also Vol. II., p. 11.)

[3] There can be little doubt that some of these Central European birds
winter in latitudes north of their summer homes.

as a centre after passage across St George's Channel.
Vast numbers of Starlings pour into Ireland by this
route between the latter half of October and the middle
of November, the passage on some occasions lasting for
several successive days.[1]   Occasionally at the stations at
the mouth of the Channel and at the Varne lightship,
in the Straits of Dover, Starlings and Rooks are recorded
as proceeding north-north-west, and as coming from
the coast of France.

Autumn Immigration from North-Western Europe.
—The arrival on our shores in the autumn of the
Starlings which make their summer homes in Scandinavia
does not begin until about two weeks after the first
appearance on the coast of England of the emigrants
from Central Europe.

The earliest immigrants from the north appear on
the coast of Great Britain during the last days of
September or the first half of October,[2] and the main
body arrives late in the latter month.   There are also
important inpourings during the early part of November,
and in some seasons laggards have made their appearance as late as the 21st of the month.[3]   A pronounced feature of these movements is that the birds
arrive in a series of "rushes," there being usually little
immigration of a straggling nature chronicled.

During these movements Starlings are recorded as
arriving on the east coast from Shetland to and some-

[1] In 1884 it was observed for eight consecutive days (15th to 22nd October)
at light-stations off the coasts of Waterford, Wexford, and Wicklow.

[2] The earliest dates recorded are as follows:—28th September 1908
(Fair Isle), 1st October 1886 and 1907, 3rd October 1884, 6th October 1883,
9th October 1882.

[3] Professor Collett informs me that most of the Starlings leave southern
Norway in the course of October, and are common at the lighthouses during
that month and the early part of November.

times perhaps beyond the Humber.   A number, too, reach the Atlantic seaboard and the Hebrides, occurring not unfrequently as far west as Sule Skerry and the Flannan and Monach Isles.

Like other visitors from the north, these immigrant Starlings appear on our shores during the late hours of the night and early hours of the morning; the other species arriving simultaneously being Redwings, Field-fares, Song-Thrushes, Blackbirds, Ring-Ouzels, Wheat-ears, Hedge-Sparrows, Redbreasts, Wrens, Goldcrests, Redstarts, Bramblings, Siskins, Chaffinches, Skylarks, Short-eared Owls, Snipe, and Woodcock.

These autumnal immigrations from the north-east are followed by overland movements westwards and southwards in search of winter quarters within the British area: the western, southern, and south-western districts of England, the Hebrides and other western isles, and Ireland affording specially favoured haunts. Ireland is entered from the north and north-east, the birds travelling by way of the Hebrides and the west coast of Scotland, or from the Galloway coast, some of them after an overland flight across northern Britain.

Autumn Passage from Northern and Central to Southern Europe.—Vast numbers of the Starlings which arrive on our shores in the autumn from both Northern and Central Europe do not remain to winter with us, but proceed on passage to retreats in South-Western Europe.

These passage movements follow (probably at once in the case of the majority of the migrants) the arrival from the Continent, and are in progress from the latter half of September (on the part of the Central European birds) until the third week of November.   The course of the birds from the east (Central Europe) has already been traced along the south coast of England and across the Channel. The birds of passage of northern origin proceed southwards by both the east and west coast-lines (including the Hebrides), but more especially the former, and finally depart as emigrants, crossing the Channel at various points between Kent and Scilly. As already stated, it is possible that some of our British Starlings may also participate in these emigrations by joining the ranks of the Continental birds and departing with them for the south.

Some idea of the magnitude of these movements may be gathered from the fact that on the night of 12th and the early morning of 13th October 1901, vast numbers of Starlings, evidently of Continental origin, passed the Eddystone, going southwards, for ten hours and a half without a break. Sixty-seven perished at the lantern, and great numbers, after striking, fell over into the sea and were drowned. Some of these autumnal visitors belong to a race which is characterised by having a purple head and throat and green ear-coverts. This form occurs on our south-eastern and southern coasts, and, as I have failed to match them with British and Scandinavian specimens obtained at the same season, I think it is probable that these birds come to us from the east.[1]

During the autumnal migratory movements Starlings sometimes considerably overshoot our western limits, and are observed far out in the Atlantic. At the end of October 1870 a large flock was encountered 300 miles west of Scilly,[2] and on 23rd October 1876 one alighted

---

[1] I captured a bird of this race on the Kentish Knock lightship. It came on board from the east in an exhausted state.

[2] Rodd, *Birds of Cornwall*, p. 292.

on H.M.S. *Alert* between capes Farewell and Clear, when 517 miles from the latter.[1] At Eagle Island, off Mayo, on 31st October and 1st November 1886, several thousands are said to have passed westwards over the Atlantic.

Winter Movements.—The winter movements of the Starling are attributable to the same cause, and are performed under the same conditions as those undertaken by the Song-Thrush, Skylark, and Lapwing. These have been fully treated in the summaries on these species, and the subject generally in the chapter (VII.) on Winter Movements. It is therefore only necessary to touch somewhat briefly on these forced migrations of this bird.

Although belonging to a species which is much affected by severe weather, and especially snow, inasmuch as its ordinary food then becomes difficult and sometimes impossible to procure, yet many of our resident Starlings remain in their accustomed haunts throughout periods of such extreme severity that great numbers perish from hunger. Others, along with species similarly affected, move to the coast, especially the west and south-west coasts of England and Ireland. Ireland is also sought by considerable numbers of emigrants, which arrive from the north-east and east on the occasion of each great outburst of cold in Great Britain. But even on the south-west coast of Ireland, where the climatic conditions are more favourable than elsewhere within our area, great numbers perish in severe seasons, such as those of 1881 (January), 1882 (December), and 1895 (January to March). Many, too, cross the English Channel and proceed southwards in search of more genial haunts on the Continent.

[1] Feilden, *Zoologist*, 1877, p. 469.

I am of opinion that these migrants are chiefly composed of our winter guests from the Continent, for careful observations made during seasons of exceptional severity, lead me to believe that most of our resident stock do not leave their usual haunts, but may be seen daily on the approach of dusk proceeding in numbers to their usual winter roosts.

Spring Movements of British Starlings.—These are return migrations of the Starlings which have wintered in our isles to their British nesting-homes.   These take place in February and the early days of March, and do not call for further notice.

Spring Immigration from Southern Europe and Passage to Northern and Central Europe.—The spring immigrations of the Starling include the return of (1) British summer visitors, of (2) the birds of passage on their way north and east from their accustomed winter quarters in South-Western Europe, and of (3) the refugees which have been forced to flee our country through the pressure of winter conditions.

The first Starlings to appear on the southern coast-line of England are probably those birds which quitted our shores earliest in the autumn—namely, the British summer visitors, which return to their breeding-haunts about the time that the first of the spring immigrants arrive on the south coast—*i.e.*, usually during the last week in February.[1]   These return movements

---

[1] The earliest record is for 19th February 1903, when great numbers passed the Eddystone in flocks, coming from the south and south-south-east.   They began to arrive at 7 P.M., and the passage lasted, with breaks, until 5 A.M.   Many were killed at the lantern, and great numbers struck and fell over into the sea.   The other species participating in this great return movement were Mistle-Thrushes, Song-Thrushes, Skylarks, Lapwings, and others not identified.

from winter retreats continue at intervals during March and the early part of April, the 12th being the latest date on which they have been chronicled. The later migrants are, without doubt, birds of passage, which after arrival proceed along both the east and west coasts (mainly the former), en route for summer quarters in Northern and Central Europe. They appear on the south coast during the night and early morning, and travel in company with Redwings, Ring-Ouzels, Wheatears, Redstarts, Blackcaps, Chiffchaffs, Willow-Warblers, and Swallows.[1]

Starlings have been noted as spring immigrants on the south-east coast of Ireland at dates ranging from the third week of February to mid-April. This indicates a return either of Starlings which have quitted Ireland for the winter, or of birds of passage on their way north ; or, again, most probably of both, for the period is wide-ranging—sufficiently so to cover both the return of native birds and the movements of birds of passage. During the later dates, these Irish immigrants are sometimes accompanied by various summer visitors and birds proceeding further north—Wheatears, Ring - Ouzels, Redwings, etc. Similar movements in the Hebrides are recorded as late as 14th April.

[1] On some occasions Starlings and other species (Skylarks, " Black Crows," Rooks, Goldcrests, and Wild Ducks) have been recorded as arriving on the south-east coast of England in the spring. Thus at the Leman and Ower lightship, forty-eight miles east-north-east of Cromer, from 11th February to 8th May 1883, the birds named are scheduled as proceeding in a westerly direction. In the *Zoologist* for 1870 (p. 2140) it is recorded from Aldeburgh that during the second week of March immense numbers of Rooks and Starlings were almost constantly arriving "from over the sea." In the same journal for 1902 (p. 87) Mr Gurney states that on 23rd March 1901 some were picked up dead on the beach at Yarmouth, along with Rooks "which had lost their lives in crossing." Similar but more regularly recorded movements are performed by the Rook.

Spring Emigration to Central Europe.—The spring emigration from the coast of south-eastern England eastwards across the North Sea of the Starlings which are returning to summer quarters in Central Europe, after wintering in the British Isles and in South-Western Europe (the latter being birds of passage), is very little in evidence as compared with the great immigratory movements on the part of these same birds during the autumn.

It comes under observation, however, in passing over the great fleet of lightships stationed between the Wash and the mouth of the Thames, and takes place between the middle of February and the end of March. There are no April movements chronicled, nor have other species been recorded as emigrating along with the Starlings. The observations on these return movements relate to the daytime only, though many doubtless pass unnoticed during the night.

Spring Emigration to North-Western Europe.—The return movements to their summer haunts in Scandinavia of those Starlings which have wintered in the British Isles, or have traversed our shores on their way from winter quarters in South-Western Europe, like all emigratory movements, come but little under notice. They are performed at night, and under favourable weather conditions, during March and April,[1] and are observed chiefly at stations on the north-east coast of Great Britain, and in the Orkneys and Shetlands where many break their journey ; the other birds noted as emigrating at the same time being Skylarks, Fieldfares,

[1] Professor Collet informs me that Starlings arrive singly in southern Norway about the middle of March, and in flocks at the beginning of April.

Blackbirds, Goldcrests, and Lapwings. The latest record was chronicled at the Isle of May on 28th April 1886, when at 10 P.M. Starlings appeared during a "rush" of migrants (Wheatears, Redstarts, Whitethroats, etc.).

The Starlings which winter in Ireland begin to emigrate about the middle of February, and in some seasons the movements are in progress until the middle or end of March. Those that winter in western Britain and certain of the Hebridean Islands (such as Tiree), leave at dates ranging from the middle of February to the end of March.

Summary of the Migrations of the Starling.—The various movements of the Starling may be conveniently summarised as follows :—

1. In June, sometimes early in the month, the young of our native Starlings gather together and lead a roving life, during which they visit the coast and other districts.

2. Later in summer both old and young form flocks and wander afield in search of food, and in the autumn many of these wanderers, notably those inhabiting the more northern and elevated districts of the mainland, seek winter quarters in the west and south of Great Britain and Ireland, some numbers of the British birds emigrating to Ireland for that purpose.

3. A portion of our native Starlings, namely, those which belong to the migratory race passing the summer with us, quit our shores in the late summer and early autumn, to winter in South-Western Europe, etc. They are essentially summer visitors to the British Isles.

4. During the autumn (late September to early November) vast numbers of Starlings arrive on the south-east coast of England from Central Europe, many

to winter in England and Ireland, others to proceed, as birds of passage, to South-Western Europe for the cold season.

5. Later in the autumn (October and November) considerable numbers of immigrants from Scandinavia arrive on our northern and north-eastern shores, many of which winter in Great Britain and Ireland, while others proceed on passage to winter in Southern Europe.

6. During these autumnal movements Starlings sometimes overshoot our western limits, and are observed far out in the Atlantic.

7. On the advent of severe cold the would-be winter residents (chiefly our Continental guests) fly to the southern and western districts (especially the coasts) of Great Britain and Ireland, and in winters or shorter periods of exceptional severity many quit our isles for more southern asylums on the Continent.

8. In February the birds which have wintered in other portions of the British area begin to return to their summer quarters.

9. The earliest days of spring, and even the latest of winter (February and March), witness the return from their winter quarters in Southern Europe of the Starlings which are summer visitors to the British Isles.

10. About the same time the refugees which were forced to quit our isles during the winter also return to our shores.

11. Early in spring, too (during mid-February and March), the Central European birds which have wintered with us depart eastwards for their summer homes on the Continent.

12. Later (in March and during April), the Scandi-

navian birds which have passed the winter in our islands take their departure for their northern summer haunts.

13. Finally (in March and April), the birds of passage, which also wintered in Southern Europe, arrive on the south coast, to travel by way of our shores to their breeding-haunts in Central and North-Western Europe.

CHAPTER XVI

THE MIGRATIONS OF THE ROOK, *CORVUS FRUGILEGUS*

THE Rook is a summer visitor to North - Western Europe, and is migratory to a considerable extent in the Central portions of the Continent. From both these areas the bird seeks Great Britain in the autumn as a winter retreat, departing again in the spring. Some Rooks—the members of a migratory British race—leave England, departing from the southern shores in the autumn, and though such emigrations or passages are somewhat scantily recorded, yet the corresponding return migrations in the spring are regularly chronicled. A similar spring immigration is also observed on the south-east coast of Ireland. The above-mentioned movements constitute the *regular* migrations of the Rook as observed in Great Britain and Ireland.

In addition, some irregular migrations and inter-migrations come under notice, for the bird is much given to wandering, especially after the close of the breeding season and during the summer, when flocks consisting of old and young visit the vicinity of the coast and some of the neighbouring islands, food of a particular nature being, presumably, the main incentive to these roving movements. In the autumn and winter there are daily movements between feeding-areas and roosting-places, these in a number of cases being situated some miles apart.

In Ireland, with the exception of the spring immigration already mentioned, the movements are, according to our present knowledge, to be regarded as being only of a partial or irregular nature.

In severe winters, Rooks, in small numbers, have been recorded as seeking certain of the Outer Hebrides in search of more genial quarters than those afforded by the mainland. To others of these islands it is a regular winter visitor, and has been known to appear at this season at remote St Kilda.

Although one of our most familiar birds—a species known to all observers—yet there is a lack of information regarding its movements that is not a little surprising—further and striking proof of the great difficulties which enshroud the whole subject of bird-migration.

Autumn Immigration from Central Europe.—This is by far the most important of the autumn migrations of the Rook witnessed on our shores, for it is from Central Europe that we receive the great majority of the foreign birds of this species which winter in Britain.

The immigrants arrive on the south-east coast of England, from the Humber to the coast of Kent, at dates ranging from the latter half of September to mid-November,[1] the greatest numbers appearing during late October, when these movements are often in progress for several successive days, during which vast numbers pour in upon our shores.[2]

The direction of the flight varies, being as a rule directly from east to west at or about the mouth of the

[1] The first recorded appearance is 16th September, in 1880.
[2] In October 1884 the migrations covered twenty-two days.

Thames (and sometimes on the coasts of Norfolk and Kent) to north-west and north-north-west on the coast of Suffolk and northwards. On reaching our shores the immigrants proceed inland in search of winter quarters.

The movements are observed during the daytime (usually between 9 A.M. and 4 P.M.) and at night; and the birds pass the lightships in straggling flocks or sometimes in small parties (even of two or three individuals), and frequently immense numbers pass in a single day. I observed a number at the lantern of the Kentish Knock lightship on the night of 17th October 1903. They were anything but sprightly in their actions, and flapped about the light in a most ungainly manner.

Mr Arthur Patterson informs me that at the Outer Dowsing lightship, off the Norfolk coast, for two or three days early in November 1902, Rooks, Jackdaws, and Grey Crows simply swarmed. One of the crew was confident that there were over a thousand birds on board at once. They were crowded on every available perchhold, bulwarks, cabin tops, every rope and fitting that was not quite vertical, nor refused ever so precarious a foothold. The weather was foggy. For two whole nights the ship swarmed with sleeping birds, and the deck in the morning was in a filthy condition.

The most frequent companion of the Rook on these occasions is the Daw, though always in smaller numbers than its congener, the other species also migrating at the same time being Grey Crows, Starlings, Skylarks, Chaffinches, and Tree-Sparrows.

Mr Caton Haigh, who is favourably situated on the north coast of Lincolnshire for observing the incoming of the right wing of these immigrants, remarks that the

parties sometimes consist entirely of old birds, sometimes of old and young, and sometimes, so far as he has
been able to determine, wholly of young birds.

In addition numbers are occasionally observed off
the mouth of the Thames and the east coast of Kent
crossing the Straits of Dover, as if coming from the
south-east coast of France.

Autumn Immigration from North-Western Europe.—
The arrivals on our shores from Northern Europe are far
from being extensive, which is not surprising, since
Professor Collett informs me that the Rook is not an
abundant species in Norway : it appears in the Shetlands
and at some of the Orkneys (including the far-outlying
islet of Sule Skerry) during October and onwards to mid-
November   The birds arrive during the night, sometimes in fairly large flocks, and often remain for a short
period before proceeding southwards.[1]

On the east coast of the mainland of Great Britain
the arrival of these northern immigrants does not seem
to have been observed ; but passage movements southwards, performed during the daytime, are recorded as far
south as Flamborough Head.   Similar migrations are
witnessed on the west coast of Scotland, chiefly at the
Hebridean stations.   These diurnal migrations are probably passage movements to British winter quarters,
and they sometimes extend as far westward as the
Flannan and Monach Isles and St Kilda.   The Rook
is a winter visitor to Barra, and probably to some
other of the Hebrides.

[1] Mr Thomas Henderson, junior, of Dunrossness, tells me that during
long-continued southerly gales he has often seen the immigrant Rooks rise
in a flock to a considerable height, as if anxious to be off, and then settle
down again. They leave Shetland for the south as soon as favourable
conditions set in.

The autumn immigrants from both East and North
settle down for the winter in Great Britain—chiefly,
I believe, in eastern England—and a few may proceed
south of the British area after arrival on our shores.

Autumn Emigration from Britain.—At the Goodwin
lightships, on several occasions during September and
October,[1] Rooks, sometimes in considerable numbers, have
been recorded as crossing the Straits of Dover in the
daytime, in an easterly and south-easterly direction, as if
proceeding to the coasts of Belgium and France. These
records are of considerable interest when considered in
connection with the more regularly observed return
movement, which occurs in the spring. The early date
at which some of these migrations are chronicled would
seem to indicate that the emigrants are British birds,
for they set in prior to the arrival of the earliest autumn
visitors from the Continent; hence the Rook is possibly
a Summer Visitor to Britain.

Spring Immigration to Britain.—During late February, throughout March, and sometimes in the first half
of April,[2] considerable numbers of Rooks, occasionally
accompanied by Daws, Starlings, and Skylarks, arrive
during the daytime on the south-east coast of England
between Kent and Norfolk, the immigrations on some
occasions lasting for several successive days.

The late Sir Edward Newton made a number of
interesting observations on these movements as
witnessed by him at Lowestoft. He writes thus of one
of them, which occurred on 31st March 1889:—" This
morning, while sitting in the house, I heard Rooks and

---

[1] The earliest of these autumn departures is dated 9th September, and
the latest 30th October.

[2] The earliest record is for 23rd February, and the latest for 18th April.

Jackdaws. On looking out I saw flocks of about one hundred coming in very high from the south-east. A few minutes later I again heard Rooks and Jackdaws, and again saw another flock, also very high, flying northwards; they were occasionally toying and circling, as one sees them in summer and autumn."

These, or perhaps we should say some of them, are, no doubt, the return movements to British haunts of the emigrants mentioned as leaving our shores in the autumn. Other individuals, especially the late arrivals, may be on passage to Scandinavia; the corresponding autumn passage southwards of such foreign immigrants is not obviously recorded in our data, though it probably occurs.

Spring Emigration to Central Europe. — As the movements in the reverse direction were the main ones of the autumn, so are these the most important ones of the spring.

As the Rooks from Central Europe were the first to arrive in the autumn, these same birds are the first in the spring to quit our shores, after wintering with us. As early as the second week of February (the 10th being the earliest record) these great emigrations from south-eastern England, eastwards across the North Sea set in, reach their maximum during March, and are much in evidence until the middle of April, the 23rd of that month marking their extreme limit in recorded observations. During this prolonged period, vast numbers of emigrants are observed at the lightships between the Humber and the mouth of the Thames (occasionally at the Straits of Dover), passing to the south-east and east during the daytime, from 6 A.M. onwards, and sometimes flying very high, Grey Crows, Daws, Sky-

I.   S

larks, Tree-Sparrows, and Chaffinches not unfrequently departing at the same time.

Prior to their departure, numbers of these emigrants have been observed passing southwards, occasionally accompanied by Grey Crows, on both the Yorkshire and Norfolk coasts, en route for particular points of embarkation, whence the passage of the North Sea[1] is made.

Spring Emigration to North-Western Europe.—The Rooks from Scandinavia which have wintered in our islands return north in March, April, and the first half of May ; and (as in the autumn) are mainly observed on passage in the Orkneys and Shetlands, including Sule Skerry and Fair Isle.   Some appear in these northern islands as early as the first days of March, but the chief movements take place during its latter days and the early days of April.   Their movements in 1904 seem to have been an exception to the rule, for I am informed by Dr Edmondston Saxby that a small flock arrived in Unst on 14th February.   Their numbers were greatly increased on 23rd March, and on 20th April enormous numbers were present over the whole island, some of which remained until 10th May.[2]   These travellers arrive during the night occasionally in large flocks, and are sometimes accompanied by Grey Crows and Daws.   The emigrants appear at stations widely scattered over both Orkney and Shetland, and usually tarry for a few days before proceeding northwards.

[1] At Somerton, on the Norfolk coast, on 20th March 1886, Rooks were flying due south in a continuous stream from 10.30 A.M. to 6 P.M., never fewer than 1000 being in sight at the same time.

[2] Stragglers have been observed as late as 16th May, and some of a party which arrived in Unst, the northernmost of the British Isles, on 4th March 1901, remained until 23rd July (Dr Saxby), and probably did not proceed beyond the limits of the British Isles.

There are only a few records relating to these movements northwards on the east coast of Britain, and it would seem as if they but rarely came under notice at any of the mainland stations. Rooks in small numbers are, however, observed annually in the Hebrides, including the Flannan Isles, on passage during March and April. They occur at the Faroes on passage about the same time (Andersen), and arrive in Norway during the latter part of March or beginning of April (Collett).

Irish Migrations. — The regular migrations of the Rook witnessed in Ireland are of an extremely limited nature, and consist of certain arrivals and departures in the spring. Ireland does not appear with certainty to be visited by Continental birds as a winter resort, and the movements observed there in the autumn must at present for the most part be regarded as irregular in their character. There are, however, occasional intermigrations with Great Britain.

Irish Autumn Movements.—During October and November in some years, Rooks have been recorded as arriving on the south-east coast, but these immigrations are so uncertain and unimportant as not to merit further notice at present. Such passages on the part of other species are among the best observed and most interesting of the Irish movements, and the absence of the Rook from the number of the regular immigrants presents a remarkable negative feature, especially since many species from Central Europe which winter in England find their way to Ireland (after traversing the south coast of England) by this route in considerable numbers.

Rooks have also been occasionally observed in October at the islands (Rathlin and Maidens) off the north-east coast, coming from the direction of the mainland of Scotland, and sometimes "rushes" are recorded there.

Irish Spring Movements.—The chief feature in the migrations of the rook as observed in Ireland is the *regular* spring immigration observed (during the day-time) on the south-east coast, between the latter half of March and the third week of April—the movements indicating that a corresponding autumn emigration most likely takes place, though it seems as yet to have entirely escaped notice. It is impossible to determine the precise nature of these movements. They may be those of native birds returning to their homes in Ireland, or of birds of passage traversing the Irish coast on their way northwards. We have, however, no further information concerning them, and the question of their nature must remain an open one.

There are occasional records of spring departures. These are witnessed at Copeland Island, Rathlin, and the Maidens, off the north-east coast, where occasionally Rooks have been observed moving towards Scotland in April. These movements, from the late date of their occurrence, would seem to indicate that the migrants were birds on passage for Northern Europe.

Apparently Erratic Movements to the West.—In the late autumn large numbers of Rooks have occasionally been observed moving westwards beyond the British Isles and out over the broad waters of the Atlantic, wherein many perish, and whence others, having retraced their flight, arrive in an exhausted condition on our westernmost shores.

> Quæsitisque diu terris, ubi sidere detur,
> In mare lassatis volucris vaga decidit alis.
>
> OVID, *Metam.*, I., 307-8.
>
> Long seeking land where none is to be found,
> The worn, wing-weary bird drops and is drown'd.

Perhaps the best instance on record[1] of such movements occurred in October 1893, when late in the month vast numbers (estimated at from 5000 to 6000) arrived at Scilly from the south-east, accompanied by a few Daws, and proceeded in a westerly direction. About the same time a large flight of Rooks, presumably the same birds, were met with by steamers out in the Atlantic some 300 miles west of Ireland. These misguided birds were in such an exhausted condition that some fell into the sea and were drowned, being too weak to retain their foothold on the vessel on which they had alighted. It is said that these birds avoided the outward-bound steamers, and sought only those which were approaching the land. As there was nothing unusual in the weather at the time of the birds' appearance in Scilly, they were certainly not on this occasion blown out to sea—a theory which has been advanced to explain similar flights.

Return movements from the Atlantic of considerable numbers of rooks have several times been recorded at stations on the west coast of Ireland. In 1884, between 2nd and 25th November, large numbers were arriving, either in flocks or at intervals, at Tearaght Island and at the Skelligs, off the coast of Kerry, for several days. Again, in 1887, between 21st October and 23rd

---

[1] J. H. Jenkinson, *The Field*, 3rd March 1894.

November, they appeared at the same stations, also in numbers and direct from the Atlantic. Similar movements were witnessed in 1888 and 1890, chiefly in November, at Tearaght and at Slyne Head, Galway.

In the middle of November 1893 (soon after the great movement observed at Scilly), some 4000 or 5000 appeared in the Island of Lewis, arriving in an exhausted state, and great numbers were washed ashore on the west side of the island.[1] At this time many too passed St Kilda, and great numbers perished. Some of them remained on the island until the following summer.

Summary of the Migrations of the Rook.—1. Partial and irregular movements on the part of young and old begin at the close of the nesting season and continue throughout the autumn.

2. Vast numbers of Rooks from Central Europe arrive on the south-east coast of England (coming from the east and south-east) between the latter half of September and the middle of November, to pass the winter in the eastern counties of England. This is the main autumnal movement.

3. During October and until the middle of November, emigrants from Scandinavia arrive on our northern shores and remain to winter in Great Britain. They are chiefly observed as immigrants in Shetland and Orkney, and on passage to their British retreats, on the north-east and north-west coastlines.

4. In severe winters some emigrate from the mainland of North Britain and are observed in small numbers in the western isles (Lewis, etc.).

5. Late in February, during March, and sometimes

[1] *Annals of Scot. Nat. Hist.*, pp. 149-150.

early in April, numbers of Rooks arrive on the south-east coast of England from the Continent, moving in a westerly and north-westerly direction during the day-time. These are most probably returning British emigrants, whose departure in the autumn is also duly chronicled.

6. Early in February, and until mid-April, the Central European Rooks which have wintered in England depart from the south-east coast for their summer homes. This is the most important movement of the spring.

7. Throughout March, April, and the first half of May, the winter visitors to Britain from Scandinavia are observed, chiefly in the Orkneys and Shetlands, returning to their northern summer quarters.

8. The Irish Movements are mainly of an irregular and unimportant nature, and Ireland is, possibly, not resorted to by the Continental visitors for winter quarters. In October and November in some years arrivals have been recorded on the south-east coast after passage across St George's Channel, and there are occasional arrivals from Scotland at the islands off the north-east coast. In spring there is a regular return migration witnessed on the south-east coast between the latter half of March and the third week of April, implying an unobserved autumn emigration either of native Rooks or of birds of passage, or of both. There are a few records of the return of Rooks to Scotland in the spring.

9. In the autumn of some years apparently erratic movements westwards over the Atlantic have taken place. During these many of the wanderers have been known to perish, while others have been observed returning, in an exhausted condition, to the west coast of Ireland, and to the Hebrides.

CHAPTER XVII

A MONTH ON THE EDDYSTONE : A STUDY IN EMIGRATION, AND OF THE CROSS - CHANNEL MOVEMENTS DURING AUTUMN AND SPRING.

WHY the Eddystone? It is true that no other light-house in any country, or of any time, has attained to the same degree of celebrity as the series of beacons which since the year 1699 have stood upon those lonely Cornish rocks ; but the halo of these romantic associations had nothing whatever to do with my selection of that station for the purpose of my bird-watching. I went there because it fulfilled beyond all others the conditions required for the prosecution of the special investigations I was wishful to carry out. During the preparation of the Reports submitted by me to the British Association on " Bird-Migration in Great Britain and Ireland," I was much impressed with the singularly deficient state of our knowledge relating to the conditions under which one of the most important and interesting phases of the phenomenon was performed—namely, that of emigra-tion. That this should be the case is not difficult to realise when it is remembered that emigration is the movement of all others which is performed under conditions of obscurity, since it is chiefly undertaken during the hours of darkness, and consequently entirely

PLATE IX.

[Photo: *Valentine & Sons*.

THE EDDYSTONE LIGHTHOUSE AND STUMP OF SMEATON'S TOWER AT LOW TIDE.

Both towers stand on detached rocks and are 40 yards apart.

escapes notice at the points of embarkation on the mainland. It does, however, come under observation at a few advantageously placed light-stations situated off the coast, where a mere fraction of the movements which take place are witnessed, for it is only under certain conditions of the weather that the migrants approach the beacon's light and reveal themselves to the watcher, if one there be.

I therefore determined, if possible, to spend a month in such a station for the purpose of adding to my personal experience in what has long been a favourite study, and in the belief that a trained observer, prepared to devote the whole of his time to the necessary vigils, might, even in so short a period, during the height of the migratory season, add considerably to our knowledge of these important movements.

It required but little consideration to decide that autumn was obviously the best season, that the south coast of England was the best section of our littoral on which to witness the departure movements from Britain, and that an ideal watch-tower would be one situated well out in the waters of the English Channel, for there the birds could be observed actually en route from our islands to their southern winter quarters. The famous Eddystone lighthouse offered all these advantages.

On hearing from me of my project, Professor Newton, with characteristic kindness, at once offered his valuable assistance, with the result that, through his instrumentality and that of Sir Michael Foster, my application for permission to reside in the lighthouse was forwarded to the Trinity House by, and with a strong recommendation from, the Royal Society. The request was most graciously granted by the Elder

Brethren, and I took up my residence on the Eddystone on 18th September 1901, and remained there until 19th October.

The Eddystone rocks consist of three contiguous reefs, which lie 14 miles south-west of Plymouth, and are 10 miles from the nearest point of the mainland. The central reef is the most extensive, its exposed length at low water being some 150 yards, while its jagged crest then rises about 15 feet above the sea. At high water all the rocks are either submerged or have their highest points awash. The lighthouse stands isolated at the northern extremity of the main reef, and is a massive structure, 168 feet in height. On the adjacent reef to the north, and about 40 yards distant, stands the basal portion of Smeaton's historic tower, completed in 1758 and in use down to 1882, a fitting monument to the genius of the founder of the science of modern lighthouse-engineering. The gallery, which was the scene of my vigils, and my perambulations too, for the base of the tower is submerged at all states of the tide, is 130 feet above the sea. The illuminating apparatus consists of a double series of dioptric lenses, one placed above the other, each furnished with a six-wick lamp, and develops the enormous power of 80,000 candles. In clear weather, however, only one lamp is used. The light is concentrated into twelve brilliant beams, arranged in pairs, which revolve slowly, taking three minutes to make a complete circuit.

Landing on the rock is somewhat exciting work for a novice, and is effected from a surf-boat towed out by the relief steamer for the purpose. This boat approaches the rock at low water, and anchors as near the base of the tower as the surf which eddies around it will permit.

Communication with the lighthouse is then established by means of a rope.   To this rope the person about to land clings with his hands, and places one of his feet in a loop formed for that purpose.   In this dangling fashion the intervening waters are crossed, the rope being payed out from the boat, and hauled in by the winch placed midway up the tower.   The only real difficulty about this unusual method of landing is to get nicely clear of the bow of the boat, and to avoid dropping into the water when the order "heave away" is given to the men at the winch.

Life on a rock-station has, of course, its little trials. He who would dwell there must, among other things, be prepared to share in all respects the lot of the keepers, their rations, and their dormitory.   He must also be content to be shut off from communication with the outer world until the monthly "relief" comes round, when, weather permitting, his incarceration ends and he returns to the ordinary comforts of everyday life.   There was one feature in the life on the Eddystone which was decidedly trying to an amateur, namely, the firing, every three minutes during fog or haze, of a charge of tonite, an explosive producing a terrific report which can be heard some 15 miles or more.   The keepers were able to sleep peacefully during these operations—an accomplishment I did not succeed in acquiring.   I may say at once, however, that the novelty of the situation, the interesting nature of my self-imposed work, and last, but not least, the great kindness of the keepers, far outbalanced those trivial discomforts which are inseparable from such a life; and I shall ever look back upon my sojourn in that lonely observatory with extreme pleasure and satisfaction.

As I anticipated, I found the Eddystone to be favourably situated for observing emigration, and, though it is probably only one among many points at which the Channel is crossed by birds en route for their southern winter retreats, yet its geographical position must be regarded as somewhat exceptional, for, in addition to many species which have sought the south coast of Cornwall from inland localities on emigration bent, hosts of migrants which have traversed the west coast of Britain from the north doubtless cross the Channel in its immediate proximity. The waters of the Channel in the longitude of the Eddystone—*i.e.*, between the easternmost part of the south coast of Cornwall and the westernmost part of Brittany—are 115 miles in width.

The amount of success which it was possible to achieve during my visit was dependent to an extraordinary degree upon the weather. This was especially the case as regards night movements; for it must be borne in mind that conditions which are eminently favourable for migration may be, and indeed in most cases are, quite unfavourable for its observation. Successful night observation I found to be entirely dependent upon certain meteorological conditions which, while favourable for emigration, also rendered the lantern attractive to the migrants—a combination which, though not very uncommon, is yet one of comparative infrequency, and this results in the great majority of movements being unobserved. The lantern of a light-station, I discovered, is simply a decoy. It is one that I found would only "work" under peculiar conditions, which are dependent upon the amount of moisture (rain, haze, cloud) present in the atmosphere. When moisture is disseminated through the air as a liquid in a state of minute subdivision, the mixture becomes more or less opaque, and the powerful beams from the lantern then become conspicuous to a very remarkable degree, and exert extraordinary attractive powers over the migrants that pass within the sphere of their influence. On such occasions the twelve slowly revolving rays from the Eddystone lantern presented a very singular and mystifying appearance, and small wonder was it that the emigrants could not resist their seductions.

My visit included a period when the nights were brilliantly moonlit and cloudless, during which, no doubt, great passage movements were performed. When such conditions prevail the bird-watcher may rest at peace in his bunk, for the migrants will speed onwards, heedless of the light and probably far above it. The rays from the lantern, brilliant though they be, are then quite inconspicuous, and the feathered voyagers pass far beyond the range of human observation. Gales were not infrequent and arrested the emigratory movements.

The first emigratory movement performed during the hours of darkness which I was to witness, set in at 3 A.M. on 23rd September, and lasted until 5 A.M. I say "set in," because just previously to its advent the weather was of such a description as to be prejudicial to, if not impossible for, migration, for a south-easterly gale was blowing with a velocity of from 40 to 48 miles an hour. Shortly before the time named, however, the wind fell to a moderate breeze, but the heavy rain still continued. Almost immediately after the wind moderated the migrants appeared in numbers. It was the first movement I had ever witnessed under such advantageous conditions, and will ever remain ineffaceably impressed on my memory. I was aroused from my sleep by the keeper on duty with the words " Birds, sir," and was on the gallery a few moments later, when the scene which presented itself was very remarkable. The birds were flying around on all sides, and those illumined by the slowly revolving beams from the lantern had the appearance of brilliant glittering objects, while the rain shot past on either hand, as I stood on the lee side, like streams of silver beads. I was not a little surprised to discover how extremely difficult it was to identify the birds seen under such novel and peculiar conditions. Even the conspicuous spots on the breasts of the Song-Thrushes were entirely effaced, as they fluttered in the beams towards the lantern, by the dazzling brilliancy of the light shining directly upon them; the smaller species had to be lifted from the lantern ere their identity could be ascertained; and the birds careering around became mere apparitions on passing from the rays into the semi-darkness beyond. A number of species undoubtedly escaped detection; but the following are known to have participated in the movement, those marked with an asterisk (on this and other occasions) having been either killed or captured :—Song-Thrushes, *Redstarts, *Sedge-Warblers, *Pied Flycatchers, *Yellow Wagtails, Turtle-Doves, Redshanks, and Curlews. The Song-Thrushes, Yellow Wagtails, and Turtle-Doves were most in evidence. The Turtle-Doves often approached the lantern, yet they recovered themselves sufficiently to avoid striking. The Yellow Wagtails killed included both adults and young.

The birds which struck the lantern did so after

travelling directly up the beams of light; but a number
of them flew high and passed over the dome.   The
emigrants came from the north and continued to arrive
and pass on until 5 A.M., but before the appearance of
dawn all had passed southwards.

This rush was evidently composed of departing
British summer visitors, spurred to move southwards by
the very unsettled weather of the previous few days.   I
witnessed no second movement of a precisely similar
nature, though, no doubt, other flittings away of our
summer birds followed; but the nights were brilliantly
fine, and the migrants passed unobserved.   On this
occasion the lantern was attractive to the birds by
reason of its rays being rendered unusually conspicuous
as they streamed out on the thick rain which prevailed.

This movement was followed by several minor
emigrations by night: that is to say, they were less
important so far as they came under observation.

On 30th September, at 9.30 P.M., following a
lifting of fog (wind E.S.E., moderate breeze, hazy),
*Song-Thrushes, *Meadow-Pipits, *Chaffinches, and
other undetermined species appeared.   The movement
ceased to be observed on the appearance of the moon
at 10.45 P.M.

1st October.—Numbers of *Meadow-Pipits passing
from 2 A.M. to 5 A.M. (wind S., moderate breeze).   At
night, on the rolling away of fog at 9.45 and during
intervals of light rain up to 11.15, *Starlings and
*Wheatears appeared at the lantern (wind S., moderate
breeze, cloudy).

10th October.—After a gale of three and a half days'
duration, the night of the 9th was clear and starlit, with
a gentle breeze from the N.N.W.; at 2 A.M. the sky

became overcast, and *Song-Thrushes, Mistle-Thrushes, Redwings, *Skylarks, *Starlings, *Meadow-Pipits, and some undetermined passerines appeared and were observed until 3.30 A.M. This was the first movement in which birds of passage were undoubtedly present— that is to say, species (the Redwing, for instance) which, having arrived in Britain from further north, had traversed our shores and were seeking more southerly winter quarters.

10*th to* 11*th October.* — During passing showers, from 7.15 P.M. to 9 P.M., Song-Thrushes and Skylarks were present. At 4 A.M., under similar conditions, several Starlings were flying round the lantern. (Wind W.S.W., light breeze.)

12*th October.* — During slight showers, between midnight and 2.30 A.M., Pipits, Starlings, and Song-Thrushes were flying in the rays. (Wind S.S.E., light breeze ; dark ; clear).

Next followed the chief movement witnessed at the Eddystone. This great southerly flight commenced at 7.15 on the night of 12th October, and continued without a break until 5.45 on the morning of the 13th. The weather was favourable for both emigration and observation. The wind was a gentle breeze from the north-east, and the very slight haze which prevailed made it necessary to burn full power in the lamps, whose rays were thus doubly brilliant, and hence extraordinarily attractive, as they streamed out upon an atmosphere eminently suited for rendering them exceptionally conspicuous.

The first migrants to appear were a few Starlings, and from 7.30 these birds were present in numbers down to almost the very close of the movement. They were

followed, in the order named, up to midnight, by
Blackbirds, Skylarks, Stonechats, Redwings, Fieldfares,
Wheatears, and Song-Thrushes. To this hour the birds
had continued to arrive and pass on in a steady stream,
while many struck the lantern. Soon after midnight a
great increase in the emigrants was observed, and the
movement assumed the character of a great rush southwards. Song-Thrushes, Redwings, Mistle-Thrushes,
Blackbirds, Starlings, and Skylarks then appeared in vast
numbers, and were followed by Chaffinches, Grey Wagtails, Goldcrests, Fieldfares, White Wagtails, Meadow-
Pipits, and Curlews. At 5 A.M. the movement was again
intensified by a fresh arrival of most of the species
named and of others, including a Grasshopper-Warbler,
which struck the lantern, while a small party of Herons
passed close over the dome, calling loudly as they flew by.

There were also many small passerines and a number
of larger birds—probably waders, from their notes—
present during the movement, but their identity was not
established. The Skylarks, Starlings, Song-Thrushes,
Redwings, and Blackbirds appeared to be the species
most numerously represented, and vast numbers of them
were observed ; but certain of the smaller birds were
almost equally plentiful. It would have been possible to
have captured some of them in great numbers ; and, as it
was, the killed or injured, which did not fall overboard,
included 76 Skylarks, 53 Starlings, 17 Blackbirds, 9 Song-
Thrushes, and examples of the Redwing, Mistle-Thrush,
Stonechat, Chaffinch, Meadow-Pipit, Grey Wagtail,
White Wagtail, Goldcrest, and Grasshopper-Warbler.

Most of the emigrants went steadily southwards, but
many, dazzled by the light, tarried, and the majority of
the species named were present in some numbers until

I.   T

the first signs of dawn, when the movement waned.   At daybreak, all, save a few Starlings resting in a dazed condition in the recesses of the windows, had passed away.

A notable and important feature of this movement was the continual arrival, down to almost its very close, of fresh emigrants, not only of the kinds early noted, but of other species which had not previously participated in it; for instance, the Meadow-Pipit did not appear upon the scene until as late as 4.50 A.M.   This continuous succession of arrivals indicated, I think, that some of the birds had come from localities comparatively near at hand on the mainland, while others had travelled from afar ere they reached the Eddystone on their flight southwards.   The presence of the Redwing and the Fieldfare showed that the wayfarers were not all natives of Britain; and it is possible that others among the migrants, perhaps the majority of them, may also have been drawn from sources beyond the limits of the British Isles.   In this connection it may be stated that all the Starlings captured at the lantern (on this and other occasions) belonged to a race having a purple head and green ear-coverts, which is said to be of Continental origin.   Regarding these Starlings, it is a fact, not perhaps without significance, that the only other specimens I have seen of this form were obtained at the Spurn Head lighthouse and at Brighton in the autumn, and were doubtless immigrants.

Throughout the movement, and especially when it was at its height in the earliest hours of the morning, the scene presented was singular in the extreme and beyond my powers of word-painting.   Hosts of glittering objects, birds resplendent, as it were, in burnished gold, were fluttering in, or crossing at all angles, the brilliant revolving beams of light.   Those which winged their way up the beams towards the lantern were innumerable, and resembled streaks of approaching light. These either struck the glass, or, recovering themselves, passed out of the ray ere the fatal focal point was reached. Those which simply crossed the rays were illumined for a moment only, and became mere spectres on passing into the gloom beyond.   Some of those that struck fell like stones from their violent contact with the glass; while others glanced off more or less injured or stunned, to perish miserably in the surf below.   Others, again, beat violently against the windows, in their wild efforts to reach the source of the all-fascinating light.   Many of those that freed themselves from the dazzling streams of light came in sharp contact with the copper dome of the tower, making it resound like a drum, and then fell like flashes into the water below, followed slowly by a cloud of feathers, resembling a miniature shower of golden flakes.   Finally, above and below the madding crowd in the illumined zone, great numbers of the migrants flitted around in all directions in the semi-darkness, and in almost weird contrast with the brilliant multitudes gyrating in the adjacent vistas of light.   The babel of tongues, too, was a very striking feature.   These were by no means the cries of enchantment, but of surprise and alarm ; and they varied from the loud rattling notes of the Blackbird and the harsh angry "churr" of the Mistle-Thrush to the faint and dainty twitter of the Goldcrest. Some Skylarks every now and then, under the impulse of excitement, no doubt, broke out into a few notes of song. Not a few strange voices were heard, some probably uttered by species with whose ordinary notes one was quite familiar ; but migrants, especially waders, have a travel-talk which is, as yet, an unknown tongue to most of us. Nor was it an easy matter promptly to assign a familiar utterance to its rightful throat, when heard under such highly peculiar conditions, and to an accompaniment supplied by the roar of the surf on the surrounding reefs.

It was interesting to note the varying degree in which the mesmeric influence of the light was exercised over the different species. The Starling was the most susceptible subject present ; and this clever bird became, under the sway of the lantern, not only a complete fool, but a seemingly willing sacrifice. It was quite fearless and indifferent to the presence of myself and the keepers on the gallery, for it hustled past us in the most unceremonious fashion to reach the lantern, and, being baulked on the threshold by the windows, made vigorous attempts to reach the seductive lamp. After having exhausted itself in these vain efforts, it sat on the sills and sashes, drinking in, as it were, the light, until it became quite stupefied, and when picked off would sit contentedly on one's hand. Great numbers were removed from the lantern and cast over into the darkness below ; but many of them immediately returned. The Skylark was nearly as frequent a victim. It came up in great numbers to the light, but not being accustomed to perch on such slight coigns of vantage as the metal framework of the windows, it fluttered violently against the glass for a time, and, becoming exhausted, sank prostrate on the gallery.[1] Tennyson has truly said :—

> . . . the beacon's blaze allures
> The bird of passage, till he madly strikes
> Against it, and beats out his weary life.

---

[1] I may here remark that I took with me to the Eddystone a quantity of netting, with which I completely surrounded the gallery by hanging it

It would have been quite possible to have captured a thousand Starlings and as many Skylarks. It was otherwise with the various species of Thrush. These, though present in equal or even greater numbers than either of the species just alluded to, were not affected to anything like the same degree. The Blackbirds and Song-Thrushes approached the lantern more freely than their congeners, but they had a habit of coming up to some extent "side on," so to speak, and consequently they glanced off either a little stunned or quite uninjured. These birds did not attempt to remain at the lantern, and those which were captured showed extreme fear. The Redwing, one of the most numerous species present, was very shy, and still more so were the Mistle-Thrushes and the Fieldfares ; the latter only approached the lantern and did not strike.

That this was a great movement is evident from the fact that the senior keeper had only once before during his sixteen years' experience seen one of equal magnitude, namely, at the Casquets, off Alderney. The other keepers had not seen anything like it before. It appears to have been a far-reaching movement, too ; for at the Bishop's Rock lighthouse, south of the Scilly Isles, and 100 miles west of the Eddystone, a considerable migration was in progress at the same time, and Starlings, Thrushes, and Fieldfares are recorded as having been captured at the lantern. It was not, however, a great night for victims apart from Starlings and Skylarks ; but had a thick drizzling rain replaced the thin veil of haze, the slaughter would, in the opinion of the keepers, have

perpendicularly from the railing. The object was to prevent any birds that struck from falling over. It answered admirably, and was the means of saving many birds which would otherwise have been drowned.

been appalling, so numerous were the emigrants and so long-continued their passage.

The bodies of the various Thrushes and Skylarks were served up at dinner for several days, and proved a most welcome relief from the tedium of salt beef, which had figured daily for some time past as the standing dish in our bill of fare.

On the night of 13th to 14th October, between 6.50 P.M. and 2 A.M., a few *Skylarks, *Starlings, *Song-Thrushes, *Chaffinches, several Turtle-Doves, and a *White Wagtail were observed at or around the lantern. The night was, on the whole, starlit and clear, but there were periods during which it was overcast, and then it was that the birds approached the lighthouse. (Wind E.S.E., gentle breeze.)

The last of the night movements during my visit was one of considerable magnitude and remarkable interest. It set in on the night of 15th October, and was in progress until nearly daybreak of the following morning. The meteorological conditions under which it was witnessed were exceptional, and afforded a clear and unmistakable demonstration of the effect of weather influences, and the extent to which we are dependent thereon for rendering the observation of migratory movements possible at such stations. In this important respect it was one of the most valuable experiences that I had. The night was bright and starlit until 7.30 P.M.; but from that hour until daybreak the state of the atmosphere was ever oscillating between intervals of brightness, and spells during which the sky was overcast and tinged with haziness, rendering full lantern power necessary. The wind was E.N.E., and varied in force from a moderate to a gentle breeze. After a little experience it became possible to tell, by watching the beams of light, what the atmospheric conditions of the moment and the chances of observation were. The beams grew conspicuous when the sky became overcast through the presence of moisture in the atmosphere, and then the birds immediately approached the beacon; but as soon as this condition passed away, the rays at once thinned down and became little more than visible, the birds sheared off, and the movement in progress ceased to be observed. During the duration of the periods favourable for observation, between 7.35 P.M. and midnight, the following species were observed :— Song-Thrushes, Mistle-Thrushes, Redwings, Skylarks, Goldcrests, Starlings (first at 10.30), Blackbirds (11.30), Wheatears (11.45), Grey Wagtails, and Stonechats (midnight). At 9.40 a number of waders passed, but their calls were in an unknown tongue. The period between 11.15 P.M. and midnight was the most productive of results. At intervals between 1 A.M. and daybreak, Wagtails, Mistle-Thrushes, Goldcrests, Starlings, Larks, Wheatears, Wrens (1.15 A.M.), Song-Thrushes, Meadow-Pipits (2.30 A.M.), Redwings, Blackbirds, and Storm-Petrels were observed—the chief periods being from 1 A.M. to 1.45 A.M., and from 2.30 A.M. to 3 A.M.; but some of the species named were observed at intervals until daylight appeared. There was practically no tarrying at the lantern, owing to the attractive periods being of short duration, and the observations afforded direct evidence that the movement was continuous and that it was in progress for at least ten hours.

The Song-Thrush and the Skylark appeared to be the most abundant species, and the latter was occasionally quite a nuisance at the lantern. The extreme scarcity of the Starling was remarkable, but, on the other hand, the abundance of the Mistle-Thrush was noticeable. The emigrants were at times very numerous, and though the atmospheric conditions were not greatly in favour of many striking the lantern, yet those killed included 11 Song-Thrushes, 8 Larks, 3 Mistle-Thrushes, 4 Blackbirds, and examples of the Meadow-Pipit, Redwing, Goldcrest, Wheatear, Grey Wagtail, Wren, and Storm-Petrel.

I will now treat of the migratory movements observed during the daytime.

It will be well to preface the observations by remarking how very difficult I found it to detect small birds at sea. This is chiefly to be accounted for by the fact that the surface of the water, being ever in motion, forms a most unsatisfactory background on which to "pick up" such birds on the wing. Dark or sober-coloured species are especially difficult to detect; but the few that showed any white in their plumage during flight came under notice almost at once.

The day migrations of land-birds observed, though of considerable importance, were entirely confined to passage movements across the Channel in a due southerly direction. The species participating in these emigrations were few, and consisted chiefly of Meadow-Pipits, several kinds of Wagtails, and Swallows; but the number of individuals of the birds named was very considerable. A few Willow-Warblers, Linnets, and House-Martins were also observed, but their numbers were so small, and the occasions on which they appeared so rare, that they do not merit further consideration.

Daily throughout my visit, when the weather was favourable—that is to say, when a light wind prevailed, no matter from what quarter—the passage of Meadow-

Pipits and Wagtails was of regular occurrence. The movements were performed during particular hours only, commencing almost immediately after daybreak—*i.e.*, from 6.15 A.M. to 7 A.M.—and were entirely over by or before midday. So rigidly were these hours adhered to by the emigrants, that I soon found the afternoons to be quite unproductive, and consequently I regulated my hours of rest accordingly.

The Meadow-Pipits often passed in small parties, consisting of as many as a score, but frequently in twos and threes, and sometimes even singly. The height of their flight varied from 20 feet, or less, above the water, to occasionally as much as 200 feet, and its direction was due south. These birds were observed on emigration, in greater or less numbers, on sixteen days,[1] during which vast numbers passed close to the lighthouse, the passage being on some days continuous between sunrise and midday. They invariably uttered their familiar notes as they flitted by. The greatest movements were chronicled on 30th September and 1st, 2nd, 3rd, 5th, and 15th October.

On the same days, with hardly an exception, and during the same hours of the morning and forenoon that the Pipits were migrating, Wagtails, singly or in pairs —but never more than three together, and that seldom —were also observed moving southwards. The species identified were the Pied, the White, and the Grey Wagtail ; but in what proportion I was unable to determine, for it was only occasionally that the birds were seen under conditions which permitted of their being identified with certainty, as they generally flew at comparatively

---

[1] I was thirty-two days on the rock, and during that period fourteen days were entirely unsuited for migration, owing to adverse weather-conditions.

considerable elevation, seldom below that of the gallery (130 feet), and most frequently over 200 feet. Sometimes, however, they broke their journey, and alighted on the reefs at low water. Wagtails were noted as emigrating on thirteen days, and, judging from the continuous nature of their passage on these occasions, great numbers crossed the Channel towards the coast of France.

Swallows were observed passing southwards on seven days ; possibly they did so on others, but they were particularly difficult birds to "pick up," even when close to the tower. On certain days (2nd and 15th October) considerable numbers passed in small parties of a dozen or so, consisting of both old and young. The movements were all timed between 7 A.M. and 11.30 A.M., and the first emigrant was noted on 24th September.

Few waders came under notice, which is not surprising, for the pelagic nature of our surroundings offered no attractions to such visitors. The most interesting of the migrants among this group was the Red-necked Phalarope, which appeared singly on two occasions, namely, on 21st September and 1st October, during unsettled weather. The first of these visitors was a bird of the year, and it remained for several hours in the vicinity of the tower, often approaching quite close to its base. The second was an adult in winter plumage, and was also under notice for a considerable time, frequently at close quarters. The 1st of October was a wild day, and the little bird was compelled to seek the lee of the lighthouse to escape the frequent squalls of wind and rain that swept past from the S.W. Both these visitors were assiduously and unceasingly engaged in the capture of some minute surface-swimming creatures, probably crustaceans, which must have been very abundant, judging from the lively actions of the Phalaropes in picking round in all directions with the greatest rapidity. They were restless, too, and constantly changed their quarters by a series of flights to try fresh areas near at hand, often, however, to return in a few moments to spots which they had just previously quitted. While thus engaged they frequently approached the edge of the reefs and did not seem to mind the buffeting they encountered amid the broken water; now and then a shower of spray would cause them to rise on the wing, but, nothing daunted, they alighted again on the water as soon as the disturbance had passed.

On 29th September a small flock of Ringed Plovers passed the lighthouse, flying rapidly due south, and evidently bent on crossing the Channel.

The Purple Sandpiper visits the reefs in the late autumn and winter, to search for food during low water, returning to the mainland at high tide, when its haunts are submerged. The first bird of the season was seen on 11th October, and as many as four were seen from that date onwards. A single Turnstone visited the rocks on 30th September—an immature specimen.

A number of migratory marine birds also came under observation. Foremost among these in point of rarity was an example of Sabine's Gull, seen near the tower on the morning of 29th September. This bird was in an interesting stage of plumage, being an adult assuming winter dress. It was most accommodating in its behaviour, since it frequently rose and displayed its deeply forked and entirely white tail, and those conspicuous bands of white which cross the pinions—features which render this species both remarkable and unmistakable when on the wing. It sat on the water more buoyantly than the other gulls around it, and was generally more elegant in form than any of them.

The next species deserving mention is the Sooty Shearwater, a bird which has no place in Rodd's *Birds of Cornwall,* and is described in the *Birds of Devonshire* as "a very rare and accidental visitor" to that county. I saw single examples on 23rd September, and on 12th (two at different times), 14th, and 19th October, the last day being that of my departure. It is possible that this bird was not very uncommon just beyond the range of identification, where the shoals of pilchards were frequent and proved a great attraction to various other species.

Great Shearwaters were very common throughout my visit, but were seen in varying numbers; on some days a few only skimming the waters around the lighthouse, while on others they were extremely abundant. When the immense shoals of pilchards were in the vicinity, I witnessed some interesting scenes in which a number of this species played a leading part, dashing into the water in the most spirited style to secure their prey; as did also multitudes of less agile Gulls of various kinds, upon whom, in turn, numerous Skuas were in close and pressing attendance. The whole formed a most animated scene—one whose interest was occasionally further heightened by the presence of a school of small cetaceans, which rolled and jumped about in all directions among the much-persecuted fish. The Manx Shearwater was frequently seen between 29th September and 14th October, but was not at all numerous.

To return to the migratory species among the

Laridæ. The occurrence of the Great Skua was chronicled on 23rd September, when three examples were observed during a south-east-by-south gale, and single birds were seen on 1st and 16th October. The Pomatorhine Skua was very abundant during the period covered by my visit, and was much in evidence when I left. Adult examples and others in melanistic plumage were not uncommon. The Arctic Skua was also common, but not nearly so numerous as the last-named species. The abundance of these piratical birds was no doubt due to the presence of vast numbers of Gulls of various kinds, and that of these last, in turn, to the great shoals of pilchards present in the neighbouring waters of the Channel.

Of the various species of Tern I saw but few examples. This was, no doubt, due to the fact that the rough water that surrounds the reefs did not afford a suitable fishing-ground. Single examples of the Sandwich Tern were seen on 25th and 27th September. A few Common Terns passed on 22nd, 23rd, 27th, and 28th September. On the morning of 12th October two Arctic Terns, in the not very commonly observed second year's plumage — the *S. portlandica* of Ridgway—came close to the tower on their way westwards.

Storm-Petrels visited us on five occasions during very unsettled weather. On 22nd September they were very abundant all around during a south-east-by-south gale, when many were engaged on the lee side of the tower in picking up food on the surface of the water, in the shape, I am inclined to think, of small particles of fatty matter from our refuse bucket. It was singular that except on these "dirty" days, no birds of this species were observed

during the daytime; but one came to the lantern at
2.30 A.M. on 16th October.

When migratory birds did not present themselves, I
found much to interest me in the habits of the Gulls,
Gannets, Shags, and Cormorants, some of which were
always present during the daytime. All the ordinary
Gulls were observed, save the Common and the Black-
headed species. I noted a fact regarding the food of
the Herring-Gull, the most abundant species, which I
do not remember to have seen mentioned, though it may
have been recorded—namely, that this bird feeds extensively on seaweed, especially on the kind known as
"sea-thongs" (*Himanthalia lorea*). Almost daily,
masses of this and other weeds drifted past on the tide,
and each patch had one or more of these gulls floating
alongside it, busily engaged in detaching suitable pieces
from the long orange-brown strings, which were swallowed
with avidity. They often squabbled among themselves
for the possession of such food-supplies. I never saw
the Lesser Black-backs, which were present in considerable numbers, pay any attention whatever to these
flotsam patches of weed.

The Gannets afforded special opportunities for observing their habits. These birds fished round the
lighthouse in numbers, and with marked success, when
the sea was rough or its surface agitated; but when the
sea was calm and its surface unruffled, they merely
passed on their way to other fishing-grounds, well know-
ing that it was useless to attempt to capture the wily
pollack, the object of their quest, when there was no
ripple on the face of the waters. The best fishing-
grounds lay at the very edge of the reefs, and hence
quite close to the tower; and thus from my elevated and

fixed point of observation on the gallery I was enabled to gauge the height from which these birds dived with a degree of accuracy not usually attainable. I witnessed many thousands of dives, but in no case did the drops exceed a height of from 130 to 140 feet. About one-fourth of the Gannets seen were in immature dress, all stages being represented except that of birds of the year.

The Eddystone was an excellent station for studying the weather conditions and their bearing upon bird-migration.

Birds when performing long flights not unfrequently pass from the zone of favourable weather, which is conducive to their departure, to an area in which the conditions are more or less unfavourable; and they are consequently recorded as arriving on our coasts in the autumn under adverse circumstances. Such inauspicious instances of immigration as these are apt to mislead those interested in the subject, for it is not always borne in mind that it is the state of the weather *at the point of departure* which affords the only true indication of the actual conditions controlling the movements.

At the Eddystone, owing to its contiguity to the mainland, one witnessed the movements and could simultaneously ascertain the meteorological conditions under which the birds elected to set out on their passage southwards. If no movements took place, either by day or by night (other conditions being favourable for the observation of night movements), then it was possible, it being the height of the emigratory season, to determine what in all probability the weather-barriers were which deterred the travellers from setting forth. Thus this station was very favourably situated—probably none more so—for observing the meteorological conditions which made for or against emigration.

No movements were witnessed, either by day or night, on the part of land-birds, under weather conditions which could be described as unfavourable for crossing the Channel.

The force of the wind is the main factor which determines what is favourable and what is unfavourable for the movements. From observation, I am convinced that the *direction* of the wind is, in itself, of no moment to the emigrants, for they flitted across the Channel southwards with winds from all quarters. It is quite the reverse, however, when its force or velocity comes to be considered, and I found that none of the movements, not even the straggling flights during the daytime, were performed when the velocity of the wind exceeded 28 miles an hour (or force 5, fresh breeze, of the Beaufort scale). With the velocity of the wind at 34 miles an hour (force 6), odd Pipits and one or two young Swallows were indeed seen in distress, and endeavoured to seek shelter at the lighthouse. The movement witnessed on the early morning of 23rd September afforded an interesting instance of the effect of the force of the wind on migration. On the wind falling from a velocity of 40 miles an hour (force 7) to 23 miles an hour (force 4), the other meteorological conditions (direction of wind and heavy rain) remaining the same, the great emigratory movement already described was initiated.

Later observations made in spring, however, demonstrated that Starlings, Thrushes, and other medium-sized species cross the Channel when the wind attains to a velocity of as much as 40 miles an hour. This, however, occurred when the birds were crossing from the south to the north, and the wind may have increased after their departure from the shores of France, or they may have passed into an area of higher velocity on approaching the English shores.

The prevalence of rain is evidently a matter of indifference to the birds. It is otherwise to the would-be observer, to whom it is most welcome, for the beams from the lantern assume additional apparent luminosity during rain, and the migrating birds are decoyed within the range of observation. On clear nights one is often dependent upon the intervention of a passing shower to learn whether migration is in progress or not, but on such occasions at the Eddystone few birds actually strike the lantern, though many fly around it.

When fog prevailed no birds were observed, though the luminosity of the rays of light then becomes most intensely conspicuous, while not penetrating beyond the immediate vicinity of the tower. During fog, charges of tonite are exploded every five minutes and produce a terrific report, which must have a decidedly scaring effect on any approaching migrants, it such there be.

The only emigratory birds observed during gales were the single examples of the Red-necked Phalarope observed on two occasions. Certain other species, such as Skuas and Storm-Petrels, the latter especially, were much in evidence when the weather was unsettled and the wind high.

An important and interesting point in connection with the phenomenon of emigration is the hour at which the departing birds set out upon their night movements. This, however, is a very difficult and obscure subject to investigate. No one, so far as I am aware, has ever witnessed the act of birds rising on the wing to depart on their nocturnal journeys[1]; while the observations made at land-stations, which may be considered to bear upon the question, are surrounded by and associated with elements of great uncertainty. At the Eddystone, and other stations situated immediately off the south coast, it seemed possible in the autumn to procure data which might enable one to fix this time of embarkation with some degree of accuracy. To this end I made a series of careful observations on the time of first appearance of emigrants at the lighthouse, and found that on a number of occasions in October this ranged from 6.50 P.M. to 7.15 P.M. On the dates on which these observations were made, the hour of sunset ranged from 5.30 P.M. to 6 P.M., but darkness did not ensue until about 6.15 P.M., or a little later. It is fair to assume that these earliest birds to appear had only a short time previously set out from localities contiguous to the shores of the mainland, some 12 miles distant. Taking these facts into account, I have come to the conclusion that when the weather conditions are favourable, the initial movement for crossing the Channel is embarked upon almost immediately after darkness prevails. On no occasion during each major movement witnessed did all the individuals of a species appear simultaneously, though sometimes several kinds arrived in company, and thus the passages were a succession of arrivals of birds previously observed. Here we have evidence, I think, that certain of the individual emigrants had journeyed from districts more or less distant ere the Channel was reached on the voyage southwards.

[1] Since my visit to the Eddystone, I have observed it at Fair Isle (see Vol. II., p. 91).

On each occasion when a number of birds of any species was killed at the lantern, it was interesting to note how considerably they varied in size, and some, though to a less degree, in colour. The Skylarks, 76 in number, obtained during the great movement of 12th to 13th October, showed the remarkable range in wing-measurement of from 4.70 in. to 3.85 in.; the Starlings, obtained on the same date and 53 in number, ranged from 5.38 in. to 4.85 in.; and the Meadow-Pipits from 3.37 in. to 2.91 in. The Skylarks and Meadow-Pipits exhibited some variation in colour, difficult to describe in words, but quite manifest to the observer. It is possible that more than one race of the two last-named species was represented during the movement, or it may be, in the case of all three species, that the variations in size, etc., were due, in a greater or lesser degree, to age or sex, or both in combination.

Wing-measurements are valuable as an indication of the range of variation within species, but speculations based upon ordinary material are apt to be extremely misleading. Here, again, sex and age, singly or in combination, may, and do, account for much of the variation to be found, and yet how insignificant are the data in our possession which afford these essential particulars!

As bearing directly upon these remarks, I will instance a few cases that came under my notice at the Eddystone. In addition to those of the Skylarks and the Meadow-Pipits (which showed a very considerable variation in size and certain diversities of plumage *inter se*, though all were obtained during a single movement), the Starlings killed on the night of 12th to 13th October were all of one race, namely, the purple-headed form, and yet the wings of the males varied from 5.38 in. to 5 in. (4 being over 5.25 in.), and those of the females from 5.15 in. to 4.85 in. (13 being over 5 in.). Some, probably most, of this remarkable variation was due to age, or individualism, none to race. This influence of age was well illustrated in the Blackbirds obtained ; the wings of the young males measured from .30 to .40 in. less than the adult. To be of any real use, beyond, of course, the important one of identification, all wings should be accompanied by the age and sex of the specimen from which they were taken, and it is important, where possible, to obtain a number of examples from the same movement. Until these essential data are forthcoming, it is impossible to realise the true significance of wing-measurements, and it is worse than useless to draw deductions from them.

As regards the characters which may distinguish the various Continental representatives of many of our commonest species, much yet remains to be learned, though, thanks to Dr Hartert, considerable progress in this branch of ornithological knowledge has been made during recent years. But the age of certain birds in the late autumn is not an easy matter to determine, for the histories of their plumages at that season do not appear to be sufficiently well known to help us to reliable conclusions on this point.

On the question of the young and old birds travelling together or apart on their migrations, or of what species follow the one practice or the other, my observations at the Eddystone throw some light. Swallows, both adults and juveniles, were observed passing in company during the daytime ; and young and old of the Mistle-Thrush, Redwing, Blackbird, Wheatear, Stonechat, Yellow Wagtail, White Wagtail, and Skylark were obtained together at the lantern at night.

Since my visit, the keepers have furnished me with a series of carefully filled-in schedules, for I succeeded in thoroughly interesting them in the work. These records, together with the wings of a great number of birds killed at the lantern, were sent to me for three years, and along with those furnished to me, as a member of the British Association Migration Committee, for the years 1884-87 inclusive, and my own observations, afford a body of information from which the following list of the birds known to visit this station has been prepared. In all, 75 species have been recorded. It is highly important to know what birds cross the Channel in the meridian of the Eddystone, the nature of their movements, and the dates on which they are performed. It must be borne in mind, however, that the Eddystone is a watch-tower pure and simple, and not an island on which birds can alight and rest for a time. Consequently, all are visitors to the lighthouse, 90 per cent. of them during the hours of darkness. This being the case, it must be remembered that it is only under particular weather conditions that the migrants approach the light, and these being the exception and not the rule, the vast majority of the migrants pass unnoticed. Consequently, a number of years' observations are essential ere a correct conception can be formed of what bird-migration actually takes place at such a station.

STURNUS VULGARIS, *Starling.*—Observed on passage in spring, autumn, and winter; chiefly at night.

In spring returns northwards from the third week in February, throughout March and April, until sometimes

as late as the first week in May. The earlier movements are those of returning British birds, the later
ones of birds traversing our shores on passage to the
north and east.

In autumn a few are observed in July, from the
second week onwards. It is frequent in its visits in
August, but the emigratory movements do not commence
in earnest until September, and, along with those of the
passage birds, last until the second week in November.

In winter it is noted moving south and west during
cold periods in November, January, and February.

FRINGILLA CŒLEBS, *Chaffinch.* — Has only been
observed as an autumn emigrant participating in the
great night rushes southwards. The dates of its
recorded appearances range from 30th September to
7th November.

FRINGILLA MONTIFRINGILLA, *Brambling.*—There are
several instances of its occurrence and occasional capture
during the spring and autumn passage movements.

In spring its appearances date from 19th April to
the end of the month ; and in autumn from 27th October
to mid-November.

CHLORIS CHLORIS, *Greenfinch.* — Only recorded for
the autumn of 1904, when single specimens were
captured at the lantern on 2nd and 20th October.

CARDUELIS CARDUELIS, *Goldfinch.* — A rare visitor
in autumn, when it has been captured at the light in
October and early November—wings sent.

ACANTHIS CANNABINA, *Linnet.*—Has been noted on
a few occasions in spring and autumn. The records are
for 5th April, 24th September, and 14th and 15th
October. On the last-named dates, I saw a few passing
south at 7 A.M,

PASSER MONTANUS, *Tree-Sparrow.*—Four were seen during the night movement to the south on 8th November 1881.

ALAUDA ARVENSIS, *Skylark.*—One of the most frequent and numerous visitors, often occurring in vast numbers in the spring and autumn, and in some numbers passing south during cold periods in winter.

The spring migrations northwards commence during the latter half of February, continue during March and April, and sometimes last until the first week of May, the 9th being the latest record.

The autumn emigrations date from 30th July, but are not much in evidence until mid-September, thence onwards to the third week of November.

The winter emigrations may take place late in November, during December and January, and the first half of February, depending upon the severity of the season.

The records, with few exceptions, relate to movements during the night-time.

MOTACILLA LUGUBRIS, *Pied Wagtail.*—Chiefly an autumn emigrant, passing southwards in considerable numbers. There are scarcely any data for the spring return movements, but one was captured at the lantern on the morning of 18th March 1904, and sent.

The autumn departures are recorded from 10th September to 4th November. Many pass between 6 A.M. and noon, and a few visit the lantern during the night movements.

MOTACILLA ALBA, *White Wagtail.*—In spring it has only been occasionally recorded on its passage northwards during the latter half of April.

The autumn return towards winter quarters has been

observed between 30th September and 14th October. I observed it in the morning from 6 A.M. to noon, and at night captured old and young birds at the lantern during the main movements southwards.

MOTACILLA BOARULA, *Grey Wagtail.*—There are no spring observations. In the autumn it has occurred from 17th September to 6th November. I captured one at the lantern on the night of 14th October, and observed others passing south on several occasions during the forenoon.

MOTACILLA RAYI, *Yellow Wagtail.* — There are a few records only, all for the autumn passage, and at dates ranging from 23rd to 30th September. On the former date I captured an adult and a young bird at the lantern and saw many others flying around from 3 A.M. to 5 A.M.

ANTHUS PRATENSIS, *Meadow - Pipit.* — Very abundant during the spring and autumn passages. Noted in spring from 25th February to 18th April, during the night and earliest hours of the morning. The autumn emigratory movements date from 4th August. Great numbers pass southwards continually during September and October, from 6 A.M. until noon; and many also visit the lantern during the night "rushes" across the Channel. It has been observed as late as the last week of November, but may then have been driven south by an outbreak of cold weather on the mainland.

ANTHUS TRIVIALIS, *Tree-Pipit.*—Examples, killed at the lantern on 29th August 1902 and 21st September 1887, were forwarded to me for identification.

ANTHUS OBSCURUS, *Rock-Pipit.*—Has been recorded as occurring during the great autumn movements, but I have never received specimens or wings for identification.

REGULUS REGULUS, *Goldcrest.*—I have only a single record for the spring—namely, for 2nd April 1902, when one killed at the lantern was sent to me.

In autumn it has occurred from 11th September, throughout October, and as late as 25th November. Like most of the other migrants, it has never been observed during the hours of daylight.

SYLVIA SYLVIA, *Whitethroat.*—As a summer visitor to Britain, and later as a bird of passage, it has occurred from the end of April (28th, earliest record) until as late as 24th May, and is not unfrequent in its appearances at the lantern.

The autumn return movements commence at the end of August, are in progress throughout September, and have occurred on 5th October, the latest date on which specimens have been obtained and forwarded.

SYLVIA ATRICAPILLA, *Blackcap.*—There is only a single record for spring—namely, that of one killed between 12 and 2 A.M. on 12th April 1902, which was sent to me.

As an autumn emigrant and bird of passage, it has been observed from mid-September down to the first half of October. One captured on 22nd November 1886 was sent to me.

SYLVIA BORIN, *Garden-Warbler.*—Has occurred in spring from the 3rd to the 24th of May.

One killed at the lantern on 10th October 1902, and sent, is the only autumn record.

PHYLLOSCOPUS SIBILATRIX, *Wood-Warbler.* — Has only twice been detected—namely, on the 3rd and 17th May : both birds were sent.

PHYLLOSCOPUS TROCHILUS, *Willow-Warbler.* — A frequent visitor to the lantern in both spring and

I.   X

autumn. At the former season it has occurred as a British summer migrant, and as a bird of passage, from 3rd April until the second week of May.

The autumn movements southwards are recorded between 15th August and 30th September.

PHYLLOSCOPUS COLLYBITA, *Chiff-chaff.*—One killed during a rush of migrants northwards, on the early morning of 10th April 1903, was sent to me.

There are records for September, but as no birds were sent, their authenticity cannot be regarded as established.

ACROCEPHALUS SCHŒNOBÆNUS, *Sedge - Warbler.*— Common at the lantern, especially in the autumn.

In spring the earliest record (bird sent) is for 28th April, and the movements continue until the third week in May.

The autumn emigrations commence early, for it has been killed (and sent) on 1st August. The departures are in evidence throughout September and down to 6th October, when the latest capture was made.

ACROCEPHALUS STREPERUS, *Reed-Warbler.* — One killed at the lantern between 12 and 3 A.M. on the 3rd of May 1887, during a remarkable rush, and sent to me, is the only known occurrence at the Eddystone.

LOCUSTELLA NÆVIA, *Grasshopper-Warbler.*—I captured one at the lantern at 3 A.M. on 13th October; and specimens have since been sent to me on several occasions for dates ranging from the last days of September to 30th October.

LUSCINIA MEGARHYNCHUS, *Nightingale.* — One killed during a rush of migrants, between 12 and 2 A.M. on the 12th April 1902, was sent to me.

ERITHACUS RUBECULA, *Redbreast.*—I have received

two specimens killed during the spring movements. These occurred on 22nd and 23rd February.

In autumn the earliest visit noted is on 2nd September. It is not unfrequent at the lantern in October, the 20th being the latest date scheduled.

RUTICILLA PHŒNICURUS, *Redstart.* — A frequent visitor to the light during the spring and autumn movements.

The earliest date for its appearance in spring is 21st April, and the latest 3rd May.

In the autumn it has occurred from 11th September to 7th October.

RUTICILLA TITYS, *Black Redstart.*—Two specimens have been sent to me—a male captured on 5th November 1902, and an adult male, in spring plumage, killed at the lantern at 2.30 A.M. on 14th March 1904.

TURDUS VISCIVORUS, *Mistle-Thrush.* — A frequent night visitor in the spring and autumn, and less so during the winter movements.

As a spring immigrant and bird of passage, it has been observed from 19th February, throughout March, to 10th April. There is a record for 10th May 1902.

The earliest dates for its appearances in autumn are the 19th and 26th August, after which it has no place in the chronicles until 3rd October, between which date and 14th November it is common.

The winter cold-weather emigrations have been recorded as early as 24th November, and occur in December, January, and February, the latest on the 10th of the last-named month, during heavy snow.

TURDUS MUSICUS, *Song-Thrush.*—A regular spring, autumn, and winter migrant, observed only during the night-time.

I.   X 2

The earliest spring immigrants from the south are chronicled for 19th February, and the movements are in progress until 13th May, the later migrants being on passage for Northern Europe.

Southward-going emigrants have been killed late in July. There are a few records for August, and numerous ones for September, October, and down to mid-November.

The forced winter movements have been known to occur as early as 22nd November, and continue, depending on the severity of the season, until the early days of February, the 8th being the latest noted in the returns.

TURDUS ILIACUS, *Redwing.* — Common during the spring and autumn passages.

Recorded as moving northwards at intervals from 19th February until 11th May, and southwards from 2nd October to 20th November.

Occurs only during the night-time, and many are killed at the lantern—200 on 21st October 1884.

TURDUS PILARIS, *Fieldfare.* — Only chronicled for the autumn passage southwards. Earliest appearance 12th October, latest 26th November. Not often obtained at the lantern.

TURDUS MERULA, *Blackbird.*—A frequent visitor in spring and autumn, and in winter during severe weather.

Occurs in spring from the latter half of February, throughout March and April, and as late as 10th May, the earlier visitors being returning British birds, the later ones en route for Northern Europe.

The autumn emigratory movements have been recorded as early as 13th August, when birds were killed at the lantern. A few appear in September, but it is not until October and the first half of November

that the great movements south are witnessed.   There
are a few notices of visits in late November and
December when severe weather prevailed on the mainland.

Turdus torquatus, *Ring - Ouzel.* — Of regular
appearance in spring and autumn.   Noted as passing in
spring from 5th to 28th April, and in autumn as moving
southwards from 3rd August 1886 (the only record for
the month), throughout September and October, and
down to 12th November.

Saxicola œnanthe, *Wheatear.* — A visitor in
considerable numbers, and frequent during the spring
and autumn migrations.   In spring it has occurred from
15th March to 30th May, many as late as 25th May.
In autumn it returns from 1st August to 6th November,
but is not abundant after mid-October.

Saxicola leucorrhoa,[1] *Greater Wheatear.*—Of this
Greenland, Icelandic, and North - Eastern American
Wheatear, I captured an example at the lantern on the
night of 16th October, and I received others which had
been killed by striking on 24th September 1903, and
on 5th October 1902.   The wing measurements of these
birds ranged from 103 mm. to 105.5 mm.   I have had
no specimens of this form sent to me as captured during
the spring movements, though it doubtless occurs.

Pratincola rubicola, *Stonechat.* — I saw, and
captured, this species at the lantern on the nights of
12th to 13th and 14th to 16th October.   Great movements southwards were then in progress, and this species
was present in some numbers, and for some time, on
both occasions.

Pratincola rubetra, *Whinchat.* — Figures in the

[1] *Saxicola œnanthe leucorrhoa.*

schedules on several occasions for May, dating from the first week to the middle of the month. I have never, however, received a specimen or a wing of this bird.

TROGLODYTES TROGLODYTES, *Wren.* — Recorded and captured on a few occasions in the autumn, at dates ranging from 6th October to 23rd November. I saw it in small numbers during the great movement on the night of 16th October 1901, and secured an example.

MUSCICAPA GRISOLA, *Spotted Flycatcher.*—There are a few entries in the schedules for May and September, but I have no record of having received specimens or wings.

MUSCICAPA ATRICAPILLA, *Pied Flycatcher.*—Has not been observed in spring. I captured one at the lantern on the early morning of 23rd September during a rush southwards of British summer birds. There are also a few other records, all for late September.

HIRUNDO RUSTICA, *Swallow.* — Observed in spring, moving northwards, from 12th April to 6th June. Most of the records are for May, when the bird occurs abundantly down to the end of that month. At this season the visits are chiefly during the night-time.

A few appear during the latter half of July, many in September, and in October as late as the 20th. In autumn it is observed, with few exceptions, between the hours of 6.30 A.M. and 1 P.M.

CHELIDON URBICA, *Martin.*—Has rarely come under notice. An adult was flying round the tower during the afternoon of 29th September, and another at 4 P.M. on 12th October. One is recorded for the night of 29th September.

COTILE RIPARIA, *Sand-Martin.*—A bird, captured on 3rd August 1886, affords the only record of the occurrence of this species.

CUCULUS CANORUS, *Cuckoo.*—There is only a single record—namely, for 23rd May 1907, when several occurred in a rush of migrants between 1.30 and 3 A.M.

CYPSELUS APUS, *Swift.*—Appears regularly in May (earliest record, the 4th), sometimes in great numbers. Many occur in June (24th latest).

Several appeared on 5th July 1902 ; and it passes at intervals during August to 10th September.

It is chronicled for the night-time (especially the earliest hours of the morning) only.

CAPRIMULGUS EUROPÆUS, *Nightjar.*—There are two records only. One was killed at the lantern during a rush on 4th May 1887, and another at 4.30 A.M. on 10th September 1902.

ARDEA CINEREA, *Heron.* — During the great movement on the morning of 13th October 1901, I heard several, at 5 A.M., croaking loudly as they passed. On the 20th, I saw a party proceeding south at 5 P.M.

TURTUR TURTUR, *Turtle-Dove.*—Regular on passage in spring and autumn. Has occurred from the middle to the end of May, and from 25th September to 13th October. All the records are for night-time.

CREX CREX, *Corn-Crake.*—Observed at both seasons. In spring it has appeared at dates ranging from 26th April to 12th May ; and in autumn from 10th September to the end of the month. It is a night visitor only.

RALLUS AQUATICUS, *Water-Rail.*—There are several records of the occurrence of this species for the autumn ; but no specimens appear to have been captured.

CHARADRIUS PLUVIALIS (*Golden Plover*).—There are only a few records, and these refer to birds passing during May, October, and November. The dates range from 9th May in spring to 7th November in autumn.

ÆGIALITIS HIATICOLA, *Ringed Plover.*—I saw small parties passing south on two occasions—namely, on 29th September at 11 A.M., and on 16th October at 1.45 A.M. On the latter date they were in rush with other species.

VANELLUS VANELLUS, *Lapwing.* — Frequently observed as a spring, autumn, and winter visitor.

Appears on its way north in spring from 20th February, and continues to pass during March and until mid-April.

It is recorded as recrossing the Channel during late September and throughout October.

It occurs in rushes at both seasons along with other species, and during the hours of darkness only.

The forced winter emigrations, under the stimulus of severe weather, have taken place from 25th November to 13th February. On 6th December 1902, some hundreds arrived at 7.15 P.M., and the movement lasted until 5.45 A.M. (7th). On the night of the 7th they appeared again at 7.30 P.M., in still larger numbers, and were striking and dropping into the surf in hundreds. The temperature at the time was very low.

HÆMATOPUS OSTRALEGUS, *Oyster-catcher.* — There are only two records for the visits of this well-known species. On 27th August 1886 a flock was at the lantern all night, and nine birds struck and were captured; and on 14th November 1887 a number were present in a rush from 6 to 9 P.M.

PHALAROPUS HYPERBOREUS, *Red-necked Phalarope.* —As already mentioned (p. 298), I saw single birds, during the daytime, on 21st September and 8th October —the only records.

SCOLOPAX RUSTICULA, *Woodcock.*—This well-known

migrant is almost unknown at the Eddystone, which would appear not to lie in the course of the numbers crossing the Channel.   I have only two records—namely, one killed on the night of 21st September 1886, and two, also killed, at 11 p.m. on 1st November 1885.

GALLINAGO GALLINAGO, *Snipe.* — Another infrequent visitor.   In November 1884, 1885, and 1887, there are records of its appearing, or being killed, at the lantern : all for 10th November !

GALLINAGO GALLINULA, *Jack Snipe.*—There are three records only, all for 1887.   On 11th November one was caught, during a rush of birds, at 2 A.M. ; on the 16th one was killed at 3 A.M. ; and on 21st December one was killed in the early morning.

TRINGA ALPINA, *Dunlin.*—There is a single record only.   One was sent to me which had been killed during a rush on 21st September 1887.

TRINGA MARITIMA, *Purple Sandpiper.*—A few of these birds visit the reefs at low water during the autumn and winter.   In 1901 they first appeared for the season on 11th October, and in 1902 on 12th October.

TOTANUS CALIDRIS, *Redshank.*—I heard the familiar notes of this species during the considerable movement which was in progress in the early hours of 23rd September.

NUMENIUS ARQUATA, *Curlew.* — Passes in spring, autumn, and winter.   At the former season it has been noted between 14th March and 2nd May ; and later in the year from 30th July to 13th October.

On 10th February 1902, it took part in a rush southwards, during snow, from 2 to 3 A.M.

Observed (heard as a rule) chiefly at night.

STERNA CANTIACA, *Sandwich Tern.*—I observed this

species moving south during the mornings of 25th and 27th September.

STERNA MACRURA, *Arctic Tern.*—I saw two close to the tower at 9.30 A.M. on 12th October; they were in immature plumage, but were not birds of the year (see p. 301).

STERNA FLUVIATILIS, *Common Tern.*—A few appeared and remained for some time off the reefs, on 22nd, 27th, and 28th September. I saw one alight on a nasty cross sea, amid which it washed and preened its feathers in the most unconcerned fashion.

XEMA SABINI, *Sabine's Gull.*—As already related, an adult assuming winter plumage appeared at 9 A.M. on 29th September, and remained some time in the vicinity of the tower, affording ample opportunities of observing it, both on the water and on the wing.

MEGALESTRIS CATARRHACTES, *Great Skua.*—I observed this bird on three occasions during my residence in the lighthouse. During a moderate gale on 22nd September, three appeared in company and came close to the tower on several occasions. On 1st October one flew past to the south-west, and on the 16th another single bird was seen.

STERCORARIUS POMATORHINUS, *Pomatorhine Skua.*—From 19th September to the day of my departure, 19th October, this species was constantly present, in great numbers, in the vicinity of the Eddystone.

STERCORARIUS CREPIDATUS, *Arctic Skua.*—Appeared with the last-named species on 19th September, and was present when I left. It was fairly common, but not so numerous as *S. pomatorhinus.*

PUFFINUS GRAVIS, *Great Shearwater.*—Seen almost daily, sometimes in considerable numbers, from 19th

September to the date of my departure, 19th October (see p. 300).

PUFFINUS GRISEUS, *Sooty Shearwater.*—I saw single examples on 23rd September and 14th and 19th October (see p. 300).

PUFFINUS ANGLORUM, *Manx Shearwater.* — Seen between 29th September and 14th October 1901, but was not numerous.

PROCELLARIA PELAGICA, *Storm - Petrel.*—A not unfrequent daylight visitor to the neighbourhood of the lighthouse in the autumn during unsettled weather. It also visits the lantern during the night-time. The records date from 21st September to 9th November, when, in 1885, ten were killed at the light.

In addition to the migratory birds whose passage movements are treated of in the foregoing list, a number of other species are known to resort to, or visit, the vicinity of the lighthouse in the autumn or winter, but regarding their times of appearance and departure we have no data. Of these I observed the following during my visit in the autumn of 1901—the Gannet, Cormorant, Shag, Great Black - backed Gull, Lesser Black-backed Gull, Herring Gull, Kittiwake, Razorbill, and Guillemot.

END OF VOL. I.

PRINTED BY
OLIVER AND BOYD
EDINBURGH

Printed in the United States
By Bookmasters